Personal Wireless Communication With DECT and PWT

For a complete listing of the *Artech House Mobile Communications Library*, turn to the back of this book.

Personal Wireless Communication With DECT and PWT

John Phillips
Gerard Mac Namee

Artech House
Boston • London

Library of Congress Cataloging-in-Publication Data
Phillips, John, 1957–
Personal wireless communication with DECT and PWT / John Phillips, Gerard Mac Namee
p. cm. — (Artech House mobile communications library)
Includes bibliographical references and index.
ISBN 0-89006-872-0 (alk. paper)
1. Personal communication service systems—standards. 2. Digital communications—standards. I. Mac Namee, Gerard. II. Title. III. Series.
TK5103.483.P47 1998
621.384—dc21 98-30072
CIP

British Library Cataloguing in Publication Data
Phillips, John A.
Personal wireless communication with DECT and PWT — (Artech House mobile communications library)
1. Personal communication service systems 2. Wireless communication systems
3. Digital telephone systems
I. Title II. Mac Namee, Gerard
621.3'845

ISBN 0-89006-872-0

Cover design by Lynda Fishbourne

© 1998 ARTECH HOUSE, INC.
685 Canton Street
Norwood, MA 02062

All rights reserved. Printed and bound in the United States of America. No part of this book may be reproduced or utilized in any form or by any means, electronic or mechanical, including photocopying, recording, or by any information storage and retrieval system, without permission in writing from the publisher.

All terms mentioned in this book that are known to be trademarks or service marks have been appropriately capitalized. Artech House cannot attest to the accuracy of this information. Use of a term in this book should not be regarded as affecting the validity of any trademark or service mark.

International Standard Book Number: 0-89006-872-0
Library of Congress Catalog Card Number: 98-30072

10 9 8 7 6 5 4 3 2 1

Contents

Foreword

Since 1991, the year ETSI published the first of the DECT standards, DECT as a technology has progressed in many directions. The target market for DECT initially was cordless telephony communications at home and in the office. However, it was designed to be capable of much more than that; in its evolution, the technology now can be found in the United States PWT system, and in other standards worldwide, spanning an ever increasing number of applications, from wireless LANs to radio local loop systems.

ETSI members have invested significant resources in the development of DECT. Working cooperatively to develop a comprehensive set of standards, they have managed to create a successful technology, which never would have had such success had it belonged to only one company. It surely is another example of ETSI making global standards happen first in Europe.

The success and growth of DECT applications have resulted in ETSI publishing many reports on DECT, in addition to the technical standards. The popularity of DECT, however, has increased the need for a comprehensive book on DECT, one that provides in an easy-to-read format the tutorial and background information not in the normal content of standards. Therefore, it gives me great pleasure to write the foreword of what may well become the definitive book on DECT.

John Phillips is well known and respected among the radio standardization committees of ETSI. He has for many years been heavily active in the development of the DECT standards, being one of those unsung heroes without whom ETSI would never be a success. His expertise also has benefited ETSI in the fields of radio local loop technologies, broadband radio access networks, and HIPERLANs.

Karl Heinz Rosenbrock,
ETSI Director-General
Sophia Antipolis, France, September 1998

Preface

Creating the Wireless Organization

At home, you like having a big, chunky cordless telephone. It is convenient and not easily lost, which is good with a family around. If it is missing, it's likely to be in one of the teenagers' rooms (and you had better be prepared for the next phone bill).

For the office, though, that big phone is not so enticing. Certainly, any analog cordless phone will give you the freedom to wander around unfettered by the cord. You can reach the filing cabinet while still talking to your colleague without toppling a potted plant as the cord trails across your desk. Nonetheless, the analog quality you get at home really is not good enough for business; there you want the quality of a digital telephone. Yes, it has to be digital.

However, a digital phone does not need to be too small, does it? After all, you only need it around the office, and it goes back on the base station after every call to recharge, but just imagine that the radio coverage extended beyond your office. What if it extended to other offices on the same floor? or the entire building? or even the entire site where you work? Suddenly it becomes more attractive to have the telephone with you most of the time, but there is a problem—the size and the weight. Now imagine that the analog cordless telephone you have at home is digital and somewhat smaller and lighter. Imagine that it fits into the top pocket of your jacket or your shirt pocket without being so heavy that you notice the weight.

Just think: You are waiting for an important call, but now you can go to the coffee machine. Or you are on your way to a meeting in the next building, and a great thought flashes through your brain. You can telephone your inspiration to one of your team for immediate action. Furthermore, your maintenance engineers on the roof checking out a problem with the air conditioning see a small flame coming from one of the fan motors. They can call the fire department immediately and not waste time running down the stairs to find a telephone.

A digital muti-cell cordless telephone technology gives you all those new freedoms around the office. You have more fun and less hassle. Life is a little easier knowing you can pick up *your* telephone anywhere to make a call or answer one. That and much more is possible with a DECT or PWT digital cordless office system. Of course, cell phones can give you those freedoms if the in-building coverage is good enough, but often it isn't. Then there is the cost. Why should you pay air-time charges for office telephone conversations?

The primary raison d'etre for the new digital cordless systems is to free you around the office from the tyranny of telephones that are attached to a place. Why would someone want to phone your *office*—wouldn't they rather phone *you*? If you spend your time doing things outside your office, discussing new ideas with colleagues in their offices, or walking the job with your people, then a DECT or PWT telephone is for you. You can't do those things confined to your desk.

Of course, with freedom comes the need for a little discipline. You have to get into the habit of turning off the ringer during meetings. It is easy to forget that the phone is in your pocket. Because the natural instinct is to answer a ringing telephone, you must just say "sorry" to yourself and turn it off. After all, if you didn't have the pocket telephone, the phone on your desk would just ring until answered by your secretary or your voice mail, and you never would have heard it anyway while you were in the meeting.

So you should not worry about not answering the machine. Take it easy. Just because you *can* answer does not mean you *must*. There are, after all, places you visit every day where it would be unthinkable to converse with someone by phone. Thus, you learn to be considerate and relaxed enough to turn off the ringer while you are in a meeting. Your phone does have a short message service, so if a call really is important enough to drag you out of a meeting, your secretary can send a silent message to your unit's display. If the situation is critical, you can make your excuses quietly and leave.

A famous 1876 Western Union memo stated "this 'telephone' has too many shortcomings to be seriously considered as a means of communication. The device is inherently of no value to us." In spite of that prediction, the telephone has changed our business and private lives for the better. The cell phone has revolutionized our business lives while we are on the road, and the new generation of cordless telephones is now about to make us less bound to our offices and better able to do our jobs with relaxed effectiveness.

That is what this book is all about: the European DECT system and its U.S. equivalent, PWT, a new means to deliver cordless telephony and other communications around the business workplace and in many other situations. Some day, all telephones will be made that way.

Acknowledgements

The authors gratefully acknowledge that they have been authorized by ETSI to reproduce parts of the DECT standards, which ETSI publishes. They also would like to thank their partners, friends, and colleagues at both Nortel and Personal Telecommunications Systems for their support and forbearance during the book's preparation.

Part I
The Technical Principles

1

An Introduction to DECT and PWT

1.1 About This Book

Why would you want to read this book? Well, first, let us tell you what this book is about, so you can answer that question for yourself. It is about the operation and applications of the digital radio communications technology, which in Europe is called Digital Enhanced Cordless Telecommunications (DECT). The U.S. equivalent is Personal Wireless Telecommunications (PWT). These nearly identical technologies can be used to design and build a whole range of digital low-power wireless communication systems, from the residential cordless telephone to the corporate multimedia digital wireless office with interconnections to a digital cellular radio system.

This book will introduce to you in detail the way DECT and PWT operate, including their protocols and their applications. If you are a manager, you may want to read Part I, "The Technical Principles," and Part III, "Advanced Features and Applications." You will get an introduction to digital cordless technology and find out what it can do. If you are an engineer, you also might want to read Part II, "Protocols and Implementation," to find out in detail how DECT and PWT really work.

DECT and PWT are described in two standards (actually two series of standards) published by the European Telecommunications Standards Institute (ETSI) and the U.S. Telecommunications Industry Association (TIA), respectively. This book is a guide to those standards, providing background explanations, describing how the systems operate, and highlighting the differences between them.

Although this book is about DECT and PWT, for convenience, we mostly refer to both technologies simply as DECT; you may take everything we say about DECT to refer equally to PWT. Where the standards differ, we explicitly cover the variations and their impact.

Part I (Chapters 1 and 2) is for those who are not familiar with the scope, capabilities, and applications of DECT and PWT. Chapter 1 is intended primarily to cover the range of uses to which the technology can be put, with a bias toward breadth rather than depth. It does give an initial technical introduction, which will rapidly get you to the first stage in understanding the technology as well as its uses.

DECT and PWT are built on a number of technical principles that define the right way to perform basic operations, such as handset location, call setup, handover, radio link maintenance, and link termination. Understanding those principles is essential to gaining an insight about the right way to implement products. Chapter 2 expands on the introduction in Chapter 1 and provides the second step in gaining a deeper technical understanding of the technology and its application. You could say that Part I explains, in what we hope is an accessible style, how DECT and PWT systems work.

Part I introduces certain principles behind the standard that you should know if you want to design products or lead product development teams. These are some of the key ideas you will need to maximize the success of a development program. It also covers the basic architecture of DECT and PWT systems, with an explanation of the roles and functions of the various parts of a range of complete products. We also explain what parts of a system are specified, what parts are not, and what options are available.

After studying the first two chapters, you should have developed an understanding sufficient to confidently discuss DECT and PWT with senior managers and with customers.

Part II, which covers Chapters 3 through 7, brings technical coverage of the system's protocols and certain implementation issues up to the expert level. It covers a narrower range of applications but in greater depth. Here we concentrate on definitive descriptions, interpretations, and details of implementation. An ordinary residential cordless telephone is used as the model. While more ambitious applications use more of the features of the DECT protocol, every application uses the basic features of what is possibly the simplest product. A full understanding of those principles in depth is the basis for mastering the details of all advanced systems.

We emphasize, via specific advice, the means to achieve robust implementations that will keep your products operating when future enhancements are made to the standards. We also cover the issues raised by the possibility of mixing terminals from one manufacturer with base stations from another.

There are certain steps you can take to minimize the possibility of failure to interoperate and so preserve the good will of your customers and minimize the need for any expensive redesign work.

Part III (Chapters 8 and 9) covers advanced cordless features and applications. It expands on the content of Parts I and II to cover a wider range of applications, although not in as much depth as Part II, because precise details depend on specific applications. You will find out about some of the means to use DECT and PWT in more ambitious systems, including cordless private automatic branch exchanges (PABXs) with corporate networking, perhaps connected to digital cellular telephone systems. Also included is material on applications that provide data communications and the use of wireless local loop bypass systems.

Because of the special place of DECT telephony in Europe, Part III also covers the essential background for gaining type approval according to the latest European Union Directives covering DECT, electromagnetic compatibility, and all telecommunications terminal equipment. This part should set you on the road to exploiting the testing and certification mechanisms put in place by the European Commission in support of DECT.

1.2 Introducing DECT and PWT

DECT is a low-power two-way digital wireless communication system. Low RF power in this context means that the normal distance of any one handset from its associated base station may range up to about 200m. In specialized applications, there are means to extend that coverage well beyond the 1,000m mark by using directional antennas.

PWT is almost exactly the same thing as DECT. Whereas the DECT system fits in with European radio spectrum plans, the PWT system fits in with U.S. radio spectrum plans. The small but technically significant differences between them have to do mainly with the manner in which the systems share their radio spectrum. In the case of DECT systems, only other DECT systems are allowed to share the European radio frequencies, while PWT must coexist with any number of other low-power personal communication systems within the U.S. radio spectrum. That is the main reason behind the differences between them, but otherwise they can be considered to be equivalent. The systems both operate at a radio frequency of around 1900 MHz, of which more in detail later.

DECT and PWT are general digital communications systems, that is, they define a communication system framework that is not tied to a specific communications service. Having its origin in about 1987 as an application for indoor telephone communication systems (albeit systems supporting hundreds or even thousands of handsets), the framework was defined so as to support other applications, too. Mechanisms are elaborated within the framework that can carry a range of telephony and nontelephony services. Applications range from simple two-way voice telecommunications, as already mentioned, to packet-switched data services using a much higher and possibly asymmetric bandwidth. Outdoor communication systems also are possible, including radio in the local loop (RLL) service and public cordless telephone systems.

That means the framework captured in what are called the base standards, is complex and abstract. On their own, the base standards often are difficult to read and comprehend. They acquire more concrete meaning and are easier to understand when they are referred to by other standards that are called application-specific profiles. In that context, a profile is a standard that states which parts of the framework are required for a specific service. Important application profiles include the generic access profile (GAP), which applies to DECT systems delivering normal voice-bandwidth speech telephony. The U.S. equivalent is slightly different and is called the customer premises access profile (CPAP). There also are profiles for carrying integrated services digital network (ISDN) service, profiles for the interworking of DECT and PWT with the global system for mobile communication (GSM) family of digital cellular telephone systems, and profiles for a number of data services. Equipment conforming to a profile should be able to interoperate with other equipment conforming to the same profile, regardless of who manufactured it.

1.3 A First-Level Technical Introduction

You likely are familiar with ordinary analog residential cordless telephones. DECT and PWT are different from analog systems and are versatile in the traffic they can carry. This introduction covers a few of the key technical aspects and highlights the significant differences between the systems. They have many more capabilities than mentioned here. (Many of their extra capabilities are described in Chapter 8.)

1.3.1 Radio Frequencies

The ordinary analog cordless telephone technologies use frequency modulation (FM) to carry speech information over the radio spectrum and frequency-division duplexing (FDD) to separate the two-way communication. In other words, they use separate radio frequencies to carry the two speech channels, base to handset and handset to base. In Europe, the telephone sets use a radio frequency close to 46 MHz to communicate in one direction and another radio frequency, at about 1.8 MHz, to communicate in the opposite direction. The radio regulators allocate several pairs of those frequencies, and simpler telephone sets can normally use just one pair. In the United States, the two frequencies are both in the region of 47 MHz because that fits in with the U.S. radio spectrum plans.

Other analog cordless telephones in Europe, called CT1, use frequencies close to the the 900 MHz band. Those telephones, however, are considered obsolete, especially because they use a radio band that now is allocated to the GSM900 digital cellular system. The second generation of cordless telephones, CT2, are digital and also use frequencies close to 900 MHz to provide a simple but versatile technology for residential telephones and small PABX systems.

Reflecting the increasing scarcity of radio spectrum, DECT and PWT use much higher radio frequencies. The frequencies are close to 1900 MHz (Figure 1.1), although there are differences between the U.S. frequency plans and the European frequency plans.

There are actually 10 radio carriers allocated to DECT in Europe and 16 for PWT in the United States (with more if you have a special license). The systems are designed to be frequency agile, that is, they are able to use any radio carrier available as long as they can see that no other system is using it at the same time. This is called dynamic channel allocation and means that a system measures the interference level on all radio channels before using the best one. In that way, a system always selects a channel that is free of interference. If some external interference arises while a system is using a particular radio carrier, the system can detect it, usually before the user knows about it, and move automatically to a new radio channel that is free. The user does not have to do anything and sometimes does not even notice the switch.

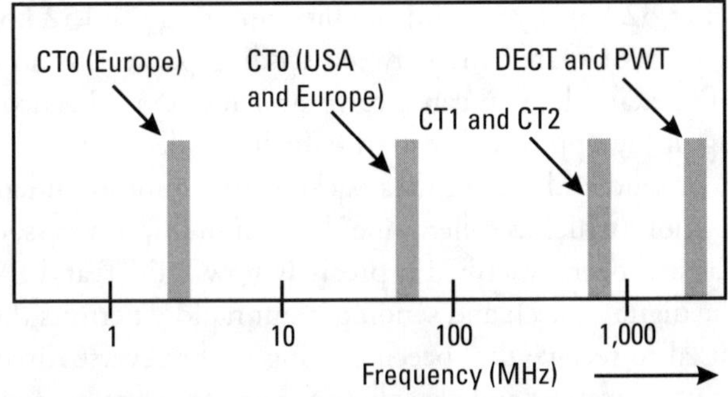

Figure 1.1 Radio frequencies used by cordless telephones.

1.3.2 Digital Communication

The next big difference between DECT and the common analog cordless telephone is that DECT uses digital communications. Of course, that's what the *D* in DECT stands for. In an analog cordless telephone, the speech is frequency modulated directly onto the radio carrier. In DECT, the user's analog speech signal is digitized using an analog-to-digital converter and digitally compressed into a 32-kbps data stream. The digital information is frequency or phase modulated onto the radio carrier for transmission and then decoded again into analog speech on reception. This is an extra stage of processing that DECT introduces between the speech signal and the radio transmission.

Clearly that extra stage adds extra cost. However, unless the radio link gets really bad, digital speech suffers less degradation in quality when transmitted over a noisy channel. Furthermore, it allows a number of important signal manipulations that ultimately save cost and enable large cordless systems to be built.

1.3.3 Packet-Mode Operation

Analog speech can be regarded as a continuous stream of information. Analog cordless telephones maintain that model, and the radio carrier in the transmitter transports the information in real time. As the information is received, it immediately is sent over the radio interface to the receiver. You also can regard digitized speech in the same way if you wish, as a stream of ones and zeroes representing the analog speech but otherwise continuous.

DECT breaks the continuous-stream paradigm. It splits the digitized speech into data packets of a certain length, normally 320 bits. Once you think of it that way, sending (or receiving) a data packet every 10 ms is almost equivalent to having a 32-kbps stream of data, but you can do more. If you take the last packet of speech data you received from the user, and instead of transmitting it at 32 kbps, you send the bits out at, say, 1,152 kbps, you find that you have a lot of spare time between sending packets to do other things (Figure 1.2). For example, you can receive similar data packets containing the speech coming in the opposite direction (which is called time-division duplexing, or TDD) or send and receive data packets from another independent telephone conversation (which is called time-division multiple access, or TDMA).

What has just been described is precisely how DECT and PWT operate. Packetizing the digital speech and sending it out rapidly permits the same radio carrier to be used to receive the speech coming in the reverse direction for any one telephone conversation and also allow the use of the radio carrier for more than one telephone conversation simultaneously. The increase in data rate from

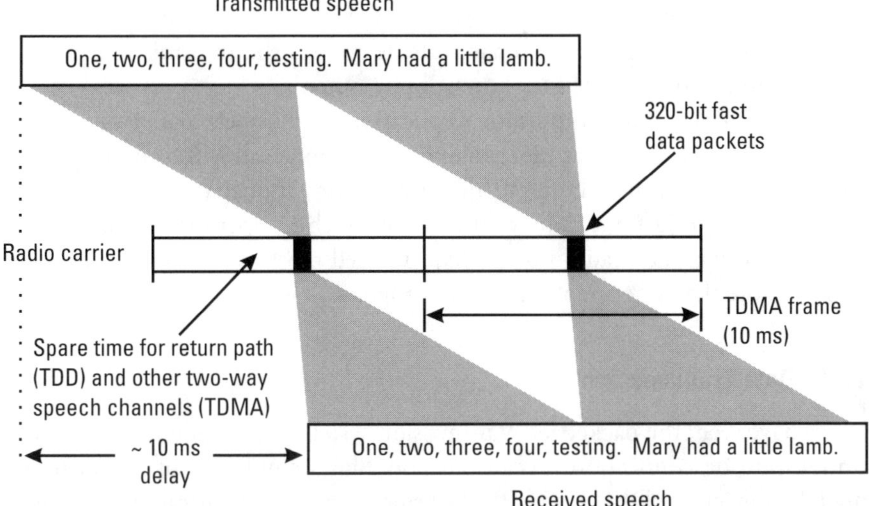

Figure 1.2 Packet-mode and TDMA/TDD operation.

32 kbps to 1,152 kbps effectively defines 24 timeslots over a 10-ms repeating period, allowing 12 two-way telephone conversations and their control and synchronization signals to be carried by a single radio transceiver. The addition of control and synchronization traffic to the user's voice traffic is why 1,152 kbps actually is more than 24 times 32 kbps. Of course, that increase in data rate and, hence, traffic-carrying capacity uses up more of the radio spectrum, so you don't get something for nothing.

The TDMA structure makes it easy to manipulate digital signals. Rather than restricting a user to only one data rate, DECT and PWT provide several different bandwidth channels by using different types of timeslots. As well as the normal slot used for 32-kbps voice transmission, there also is the half-slot, intended for use when a cost-effective low-rate (8 kbps) speech coder is adopted, and the double slot, which provides a capacity of 80 kbps (suitable for uncompressed digital telephony or ISDN). There even is a short slot for broadcasting signals and information to the handset without any user traffic.

The TDMA structure also allows cost-effective methods of increasing a system's coverage in low-density applications. It is possible to use a wireless repeater, which receives a signal on one timeslot and retransmits it on another. It also is possible to use distributed antenna techniques, which allocate different timeslots to the different antenna elements.

You will have noticed that to transmit a packet rapidly, the entire packet has to be assembled first. The same applies at the receiver, where the packet must be completely received before it can be expanded and converted back to analog speech. That adds delay into the transmission process, which must be compensated for in telephony systems.

One of the most important applications is business communications, where it is highly likely that more than one telephone call will be in progress at a radio site at one time. Handling all conversations at once (well, at least up to 12 of them) with just one radio transceiver is much cheaper than having several radio transceivers at a radio site, perhaps unused most of the time, just in case there is more than one telephone call in progress.

1.3.4 Data Transmission

As we have seen, the packetized transmission system allows emulation of a continuous analog connection. Because it also allows emulation of a continuous digital connection, DECT and PWT systems can carry continuous streams of data that are not speech.

Furthermore, such a system can carry data that is inherently in packet form, arriving at irregular and unpredictable intervals, without hogging the radio spectrum at all times just in case a data packet should arrive. Data packets for transmission simply are queued up and dropped into the first available timeslot on a radio carrier. This sort of data pattern occurs, for example, when Internet protocol traffic is being carried.

1.3.5 Cellular Operation and Mobility

If all that DECT and PWT systems could support were 12 duplex speech channels with one radio transceiver in one radio cell, they would not meet the needs of more than a small number of today's cordless office applications. With very low power consumption, and hence a radio range of about 200m, a single cell would not cover much of a medium-sized business's premises. For that reason, DECT and PWT are designed with support for multiple radio cells to provide continuous coverage over a wide area like a miniature cellular system for which you don't have to pay air-time charges (Figure 1.3).

That means DECT and PWT systems have what is called handover. Handover means you can keep talking to someone while walking around your office and moving between areas covered by different radio cells. The same basic mechanisms that allow the portable to escape from interference allow it to swap its radio link from one cell to another whenever the radio signal from one cell gets too weak and there is a better cell in the neighborhood. The process called seamless handover means you set up a second radio link to a new cell before you drop the old one. That means a cellular fixed system can arrange, at

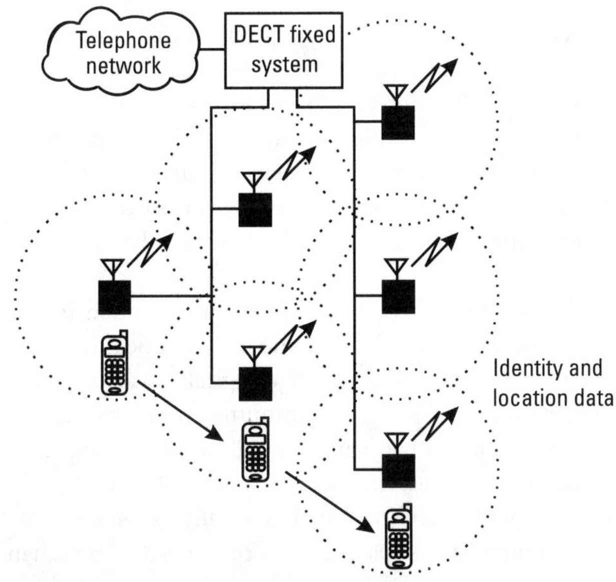

Figure 1.3 DECT and PWT support cellular operation.

the handset's request, for the conversation you are having to come simultaneously from two different cells. That way, the handset can swap from one to the other without a break before giving up the old link to the weaker base station.

Being like a miniature cellular system also implies having roaming. That means the calls coming into your handset will find you wherever you are within a multicell system. To achieve this your handset will always keep the fixed system informed of where it is. The base stations broadcast information about their system identities and can be divided into location areas, each of which broadcasts a different location area identity as well. If your handset notices that it has moved into a new location area, it will send a short message to the fixed system that says, in effect, "Hello, I'm handset 1234 and I'm now here." That way, the system can keep track in its location registers of where each handset currently is. When an incoming call arrives, it can broadcast a paging message over each cell in the right location area that says, in effect, "Call for handset number 1234; please alert the user."

1.4 Cordless Applications

As we have seen, DECT and PWT are general-purpose systems because they define a communications framework rather than a single application. The framework generally assumes that two end points want to communicate using a point-to-point service (although there are some situations in which DECT and PWT allow point-to-multipoint communications). Those end points technically are a fixed part and a portable part (often, but not always, a base station and a handset). There are special allowances for pairs of portable parts to communicate directly, but that is done by allowing one of the portables to broadcast system information that makes it look like a fixed part to the other portable.

Between the two parts, the radio spectrum is divided into a number of communications channels. Typically, one communication channel is full duplex (that is, it is two-way) and carries a digital synchronization channel, a digital signaling channel, and a digital communications channel capable of carrying encoded voice signals. That is just enough to carry a single two-way voice conversation, just as you would expect from a normal wired telephone. On the wired network, a digital speech channel normally is 64 kbps each way, but DECT and PWT squeeze 64-kbps coded voice into a 32-kbps channel, using a standard voice channel compression algorithm called adaptive differential pulse code modulation (ADPCM) [1]. In Europe, the radio spectrum is allocated to make available 120 of those two-way compressed voice channels. In the United States, there are 96 for voice and 96 for data.

The system's versatility lies in its assumption that you might want to do things differently within the same framework. First, you are allowed to concatenate channels to carry more data between a fixed part and a portable part. Instead of carrying ADPCM-coded voice signals, you could carry an ISDN basic rate service (144 kbps) spread over several DECT channels. Furthermore, the system makes the assumption that you might not want to have a symmetrical communications path at all and that you might need more bandwidth in one direction than the other.

There is more. You are allowed to use a channel discontinuously, that is, you can communicate only when you need to. That capability typically is used for data communications rather than voice and allows other communication paths to use the same radio spectrum when you are not. The range of applications is outlined here and covered more extensively in Chapter 8.

1.4.1 Residential Telephones

A residential cordless telephone system is probably the simplest possible cordless system. In many cases, it is just a single telephone handset and a single base station connecting to an external telephone line. The coverage area depends on local conditions but could be up to 200m or so in open-plan conditions, down to 50 m or so in offices with metallic partition walls.

The system is quite secure. In the first place, all handsets have to have a unique electronic serial number. That number usually is stored within a base station when a handset is registered with it, rather like a cut-down version of the digital cellular system's home location register (HLR). There may be, for example, a small push-button switch on the base station that you press to invoke the base station's subscription mode for a short period. The handset will have some registration feature that when activated contacts the local base station. If the base station is in subscription mode, it allows the systems to exchange identity codes before the base station returns to normal operation. Once the base station knows the identity of the handset (and vice versa), the two are able to communicate. The base station always rejects communication attempts from unregistered handsets.

The first level of security, therefore, lies in the protocols that require mutual recognition before the handset is allowed to use the telephone line. The physical security of the base station prevents anyone else from registering another handset with it so you cannot just take an ordinary DECT or PWT handset and use someone else's line without physical access to their base station.

1.4.2 Telephone Systems for Small Businesses

A small business telephone system may have only a few more features than a residential cordless telephone. Like a residential system, it also may have a single base station to provide radio coverage over a small shop or office. However, it might support more than one handset. Because DECT and PWT use a TDMA system to separate 12 duplex traffic channels on one radio carrier, a single radio transceiver within the base station could support up to 12 simultaneous two-way conversations over up to 12 telephone lines. The radio transceiver often can be made more cheaply if it is designed to support only six handsets, which usually is sufficient.

Also, instead of supporting six or 12 external telephone lines, a business system could provide just two lines and also provide intercom communication between handsets in different offices, for example. A single base station supporting six handsets and two telephone lines is a common product, aimed at small businesses or even busy households.

Security in a business environment may require more than the registration security of a handset and a base station. It may be necessary to ensure that the base station and the handset both support the standard encryption process, which permits a digital channel to be encrypted in such a way that within the foreseeable future nobody but a major government could decode the conversation within a reasonable period.

1.4.3 Large Business Telephone Systems

In the realm of the larger business, coverage from a single radio base station (i.e., a single radio cell) simply is not enough. A 1,000-person site containing two buildings, each with four large floors, may need 16 or more radio cells (at two per floor) to ensure that everyone will be within radio range of a base station at all times. The radio cells will be connected back to a cordless system controller, probably using simple twisted-pair cables within the building's structured wiring. The controller could either be integrated within a cordless telephone switch or be a separate adjunct system connecting to a normal PABX via digital or analog lines.

Such a cordless PABX operates very much like a miniature cellular radio system, delivering incoming calls to a handset wherever it is within any part of the radio coverage area and taking outgoing calls from it. As you move around your building, most business systems will hand over your call from radio cell to radio cell, usually seamlessly without your being aware of it.

Within the cordless controller, if it is large enough, there might be a database that keeps track of the location of all handsets. In a digital cellular system, that would be called the visitor location register (VLR). A VLR may not be necessary in smaller multicell systems because notification of an incoming call for a particular handset (which is called paging) can be delivered over every radio cell simultaneously without using up all the system's paging capacity. In larger systems with many incoming calls, it may be necessary to use a VLR to restrict paging for a specific handset to certain location areas where the handset is known to be, thus preserving the system's overall incoming call capacity.

1.4.4 Corporate Multisite Telephone Systems

There usually is some limit to the length of wiring that connects a radio cell to a cordless control system. If your business has buildings in different parts of town or in different towns, a single cordless controller will not be enough to allow people to receive their calls regardless in whichever office they are currently to be found. The protocol itself does not include support for multisite networking because it is an air interface standard, not a network standard. The protocols on the air interface, however, include support for networking of separate cordless PABXs across a corporate network. That might be via an ISDN line between the two sites, for example, running an enhanced system-to-system signaling protocol.

In that situation, the system usually is large enough for location databases to essentially allow an incoming call to be rerouted to the building or site where you are. The protocol allows handsets to recognize their location and requires them to register their presence, so location databases can be kept updated. The location databases are not part of DECT or PWT. They are an application that is supported by the cordless system, but they are not specified in the standards.

1.4.5 Data Communications

DECT and PWT are digital communication systems, based on the transmission and reception of packets of data. Hence, only one of the possible applications is for speech transmission. More generally, they support circuit-switched data transmission, rather like a modem used over the public telephone network. If you have continuous streams of data to be transmitted, then the protocol allows data streams of various data rates to be established. As we have seen, DECT and PWT also support packet-mode transmission in which packets of data are sent and received on demand rather than on a regular basis. Internet connections are just like that.

It is difficult to list all the possible applications for data transmission that DECT and PWT can handle because the packet-based transmission medium is, like asynchronous transport mode (ATM) a very general data transport mechanism.

1.4.6 Public Telecommunications Applications

Out on the street, a DECT or PWT handset can be your own private pay phone. It can be used to make and receive calls in public areas such as airports and railway stations and to provide all the other communications services that the system is able to handle.

That idea, sometimes called telepoint, has been tried at various times with limited success. Compared to cellular telephone service, the coverage areas have been rather limited because extensive public radio coverage based on a low-power cordless radio technology is more difficult to achieve. More sites need to be found and more equipment has to be deployed (even if the equipment at each site is simpler). Also, some previous attempts have implemented outgoing calls only, to save on infrastructure cost, even though the technology was capable of delivering incoming calls as well.

In spite of the disappointing history of telepoint so far, almost all cordless services can be delivered via a public infrastructure, with the secure authentication that is essential in any application for which a usage charge is made.

1.4.7 Radio Local Loop Applications

Although the materials used are fairly cheap, the copper cable from the local telephone exchange to your home, the local loop, is an expensive piece of the telephone company's plant to maintain. It goes wrong in all sorts of subtle and intermittent ways, such as partially short-circuiting when it rains. In the long run, a radio carrier often is cheaper to maintain, even though it may cost more to install. For that reason, there now are many different RLL systems, some purposely designed as local loop systems and some adapted and enhanced from systems originally intended for other applications.

DECT and PWT have been enhanced to provide good remote replication of most of the services available via a standard analog local loop. A separate standard also allows them to deliver digital telephony to a fixed ISDN terminal connected to a socket on a remote portable. However, local loop systems generally are fixed installations, so the mobility services offered by the protocol are not used. The fixed nature of the installation enables the use of high-gain antennas to permit greater-than-normal radio range to be achieved. In local loop systems, good radio range is an essential feature because the optimal positioning of base stations is much easier if the range is large and the required number of base station sites is reduced. RLL systems using DECT and PWT often are not engineered with as much link budget as purpose-designed RLL systems; nevertheless, they can provide convenient RLL service in a large number of situations.

1.4.8 Multimode Cordless and Cellular Systems

The radio coverage area from a cordless PABX system usually is smaller than the coverage area you receive with the average digital cellular system, such as GSM900, DCS1800, or PCS1900. From the PABX, you generally get coverage only while you are indoors at your office. If, however, you combine in one handset a DECT transceiver and a GSM900 transceiver (or PWT and PCS1900), you can get the best of both the cordless and the cellular worlds.

In-building, your dual-mode handset recognizes the transmissions (the beacons) from the cordless PABX. The handset sends its identity to the system and attaches to it, telling the PABX to deliver incoming calls. In this mode, you do not use cellular air time, which is likely to be cost effective. When you move out of the office, the handset cannot see the PABX any more and switches to cellular mode. If it can see a cellular control channel, it registers with that system and you can receive your incoming calls and make your outgoing calls via the cellular system.

The clever bit is when there is communication between the location registers in the cellular network and the location registers in the PABX, so that calls arriving at your own private network on your office telephone number can be redirected to the cellular system when you are out of the office and vice versa. However, unless the PABX is supplied by the cellular system operator, the owner or operator of the PABX and the cellular system operator may need to come to an agreement over significant commercial issues concerning usage charges.

1.5 A Brief History

The brief history presented here certainly is not complete, and the story of DECT and PWT is still unfolding. However, this section does chronicle some important events in developments so far.

1.5.1 European Cordless Telephones in 1987

In early 1987, the United Kingdom published British Standard BS6833 [2] for second-generation cordless telephones (CT2) based mostly on work over several years at British Telecom's research laboratory. The work was done primarily in anticipation of a rise in demand for cordless telephones and in response to the observed increase in imports of unapproved analog cordless telephones from the United States. The cordless telephones then legally in use (in the United Kingdom and in other parts of Europe) were either (1) "CT0" telephones, which were analog FM cordless telephones with a very small fixed frequency allocation, relatively poor audio quality, and questionable security, or (2) Europe's 900 MHz analog FM CT1 telephones [3], which were much better in quality (but rather expensive) and not permitted within the United Kingdom and which used part of the GSM frequency allocation. Some other countries, notably France, had tightened up their standards for analog CT0 telephones to improve their quality while retaining their low production cost. However, as you can see, at the time, the European cordless telephone market was fragmented.

1.5.2 The Birth of DECT

In late 1987, ECTEL, an association of European Union telecommunications system manufacturers, took a look at how to specify digital cordless telephones for the entire European market. They attempted to decide between TDMA and frequency-division multiple access (FDMA) systems, finally decided that

TDMA systems would be the more flexible, and formed several working groups to put together a standard.

Those working groups put together the principles that today shape DECT and PWT. The principles have proved to be so flexible that all the applications now covered by the technology have been made possible. The packet transmission and reception scheme, the number of duplex channels per radio carrier, the radio carrier frequency spacing, and many more features all were basically decided at that point, even if the radio spectrum had not at the time been officially found.

The acronym *DECT* originally stood for Digital European Cordless Telephone, and its meaning has since changed twice. It was first changed to Digital European Cordless Telecommunications, to reflect the growing number of nontelephony applications that the DECT system was incorporating. In 1996, it was changed to Digital Enhanced Cordless Telecommunications, to reflect the adoption of DECT in many non-European countries.

1.5.3 DECT as an Official European Standard

In 1988, ETSI (the European Telecommunications Standards Institute) was formed as a result of a European Commission's Green Paper on telecommunications standardization. By 1989, ETSI's Radio Equipment and Systems Technical Committee (TC/RES) had formed a subcommittee (called RES03) to incorporate the DECT working groups from ECTEL. It is that subcommittee, now called the DECT Project, that has driven the development of DECT systems and standards ever since.

In 1991, the first DECT standard was published, and by mid-1992, the first complete DECT base standards [4] were available. They did not require equipment interoperation, except in the case of handsets and base stations intended for public access use. In 1993, however, the European Commission decided that basic voice telephony was an application in which the European consumer should be able to rely on all DECT equipment interoperating. From that decision, the concept of the Generic Access Profile (GAP) was born [5], as an addition to the DECT base standards to make sure voice equipment would interoperate in public, residential, and business applications.

In mid-1996, the second edition of the DECT base standards appeared, along with the first edition of the GAP. The new base standards corrected many small technical loopholes. The first GAP-compliant DECT systems started to appear in 1997. At that point, the number of other application profiles for DECT started to increase rapidly, as data transmission and DECT/GSM and DECT/ISDN interworking became increasingly important, as well as RLL applications. Even today (early 1998), new standards are appearing, and work is going on to improve the base standards and their range of applications. You can see from the length of the bibliography at the end of this book just how many applications there are.

1.5.4 DECT Becomes PWT in the United States

In 1995, the TIA adopted a version of the DECT standards under the names Personal Wireless Telecommunications (PWT) [6], and PWT-Enhanced (PWT-E) [7], the first as an unlicensed system and the second as a licensed version. The main differences between DECT and PWT were those needed to fit DECT into the frequency spectrum available in the United States. The greatest task in achieving that was to enable PWT systems to share radio spectrum with other non-PWT applications conforming to the FCC's spectrum-sharing etiquette [8] for the personal communications system (PCS) band. However, there also were small changes to the GAP, which became the Customer Premises Access Profile (the CPAP) [9].

1.5.5 The Developing Market

DECT cordless telephones have been available in Germany since about late 1992. One of the authors bought a single-cell two-line small office system in Berlin in August 1993, at which time DECT equipment of that type was available almost nowhere else. By 1996, however, single-cell residential and small business telephones supporting up to six handsets on a one-line or two-line base station were relatively common in many European countries and had fallen in price to about half what they cost in 1993. By the start of 1998, similar small systems had almost halved in price yet again, and the range of equipment available had increased enormously.

During the period 1992 to 1996, proprietary cordless PABX systems, conforming only to the DECT coexistence standards, were on sale from a small number of sources. By late 1997, there were many more manufacturers of DECT equipment in the European market, most of whom had abandoned the original proprietary approach and had adopted the latest interoperation specifications. Several of those manufacturers also had PWT versions for the United States.

Those developments have been fueled by the availability of standard chipsets in multimillion quantities at reducing prices. From 1995 through 1997, the market for DECT chipsets, largely in Europe, doubled each year, leading to a volume of about 6 million sets in 1997. That doubling is expected to continue for the next two years at least. As a result, we surely should expect to see further developments in the worldwide marketplace in new services and in competitive pricing.

References

[1] ITU-T Recommendation G.726 (1991), *40, 32, 24, 16 kbit/s Adaptive Differential Pulse Code Modulation (ADPCM)*, Geneva, 1991.

[2] British Standards Institution, *Apparatus Using Cordless Attachments (Excluding Cellular Radio Apparatus) for Connection to Analogue Interfaces of Public Switched Telephone Networks*, BS 6833 (Parts 1, 2, and 5), London, 1987 (amended in 1992).

[3] ETSI, *Radio Equipment and Systems (RES); Technical Characteristics, Test Conditions and Methods of Measurement for Radio Aspects of Cordless Telephones CT1*, I-ETS 300 235, Sophia Antipolis, France, March 1994.

[4] ETSI, *Digital Enhanced Cordless Telecommunications (DECT); Common Interface (CI)*, ETS 300 175 (9 parts), Sophia Antipolis, France, September 1996.

[5] ETSI, *Digital Enhanced Cordless Telecommunications (DECT); Generic Access Profile (GAP)*, EN 300 444, Sophia Antipolis, France, August 1997.

[6] TIA/EIA, *Personal Wireless Telecommunications Interoperability Standard (PWT)*, TIA/EIA 662-1997.

[7] TIA/EIA, *Personal Wireless Telecommunications—Enhanced Interoperability Standard (PWT-E)*, TIA/EIA 696-1996.

[8] Federal Communications Commission, *Unlicensed Personal Communications Service Devices, CFR 47 Part 15, Subpart D.*

[9] TIA/EIA, *Personal Wireless Telecommunications Interoperability Standard (PWT); Part 9: Customer Premises Access Profile (CPAP)*, TIA/EIA 662-9-1997.

2

The Technical Principles

2.1 A Cordless Framework

The terms *DECT* and *PWT* refer primarily to access standards (i.e., "air interfaces") rather than applications. Neither covers an entire usable digital cordless communication system. Only the following parts are specified:

- The coexistence parameters and procedures of a radio interface, which ensure mutual noninterference of equipment using the same radio spectrum;
- The internal (peer-to-peer) operation of a pair of cordless link end points (called the fixed and the portable terminations), which work together to provide cordless communication services;
- The interfaces at which an application drives the protocol to obtain those services.

As can be seen from the preceding list, the radio interface is specified at two compliance levels. At the first level, it is specified in such a way as to ensure the sharing of the radio medium among different applications while operating in the same spectrum at the same location, with a minimum of mutual interference. At this compliance level, the internal communication on the radio interface is regarded as proprietary, and interoperating portable and fixed parts have to be procured as matching pairs to ensure their compatibility.

At the second compliance level, certain services (principally voice telephony in the European Union) must meet a standard for the detailed operation of the air interface. At this compliance level, the radio interface can be regarded as an extension to the telephone wall socket. Instead of just being able to plug in any normal fixed telephone, you get the ability to use any standard cordless handset. In the DECT case, users in the European Union may expect to be capable of receiving basic telephony service through any DECT portable, from any manufacturer, when registered with a suitable base station. In the United States, PWT is also able to achieve intermanufacturer interoperation for voice systems, but that capability is not mandatory.

The other essential parts of a complete cordless system, called the *application* at the portable end and the *local network* at the fixed end, are left almost entirely unspecified by the base standards. The only matters covered are some basic telephony requirements whenever the system is used for voice service. Those specifications include audio levels and echo cancellation requirements, which do not apply when the application is for nonvoice traffic.

By creating such a framework, the base standards are as independent of the application as possible. That way, the base standards are versatile and permit the design and construction of many different cordless applications, including those not foreseen today.

2.1.1 The Role of DECT or PWT in a Cordless System

Because DECT and PWT do not specify a complete cordless system, we need to know the boundaries between what is specified and the applications that are supported. That is illustrated by the example in Figure 2.1.

Take a simple residential telephone. As you can see from Figure 2.1, making the application cordless is done by splitting the equipment into two sections, the portable part (the handset) and the fixed part (the base station). The two parts are connected by radio whenever a user goes "off hook" (perhaps by pressing a "line" button), and the radio link is disconnected whenever the user hangs up. In a wired telephone system, the only interfaces that are standardized are those to the user's mouth and ear (by virtue of human anatomy and the psychology of human hearing) and the interface to the telephone network. Therefore, the means by which the radio link connects the two parts of a cordless telephone are not constrained by any telephony interfacing requirements and are constrained only by the need to maintain certain audio transmission requirements such as loudness, distortion, and delay.

We can now split the cordless telephone into smaller components, assuming that we want to use DECT or PWT to connect the fixed and portable parts. The DECT framework assumes that the two ends of the link each comprise a cordless termination (a fixed termination or a portable termination) and an application that interfaces to the termination and uses its services. The

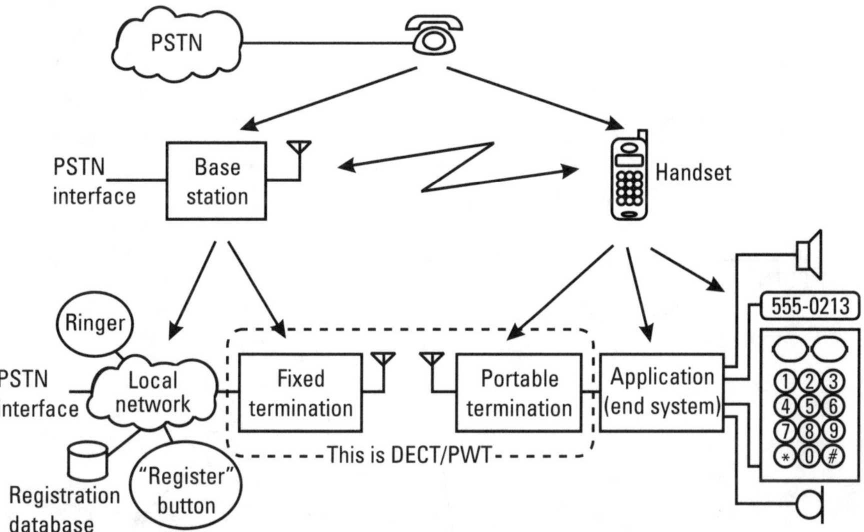

Figure 2.1 The DECT/PWT parts of a simple cordless telephone.

standards specify only the requirements for the operation of the terminations and an abstract interface to allow an application to drive them.

In this particular simple application, the portable's microphone, keypad, display, and loudspeaker (maybe a separate ringer, too) are all part of the application. In this case, the application is a telephone handset. At the residential base station, the application, called the local network by the standards, comprises an interface to the telephone network, supporting a ringer and a registration store to keep track of the identities of all the handsets that are allowed to connect to the line. The registration database is the means to deny access to unauthorized handsets. The physical security of the base station (and, in this case, its handset "register" button) is just as essential here as it is for stopping someone from using your fixed telephone or plugging their telephone into your jack and using your line.

In more complex multicell applications, for example, a cordless PABX (Figure 2.2), there may be many fixed terminations connecting many portable terminations to a network containing a switching system with mobility management, including handset location registers.

In Figure 2.2, the fixed terminations each are split into a central control fixed part (CCFP) and several radio fixed parts (RFPs). Such a split is typical in multi-cell systems, where only the essential parts of the radio transceiver are located at the radio cells. Connection from the RFPs back to the CCFP typically uses twisted pair cables, which carry signaling and user traffic, and also

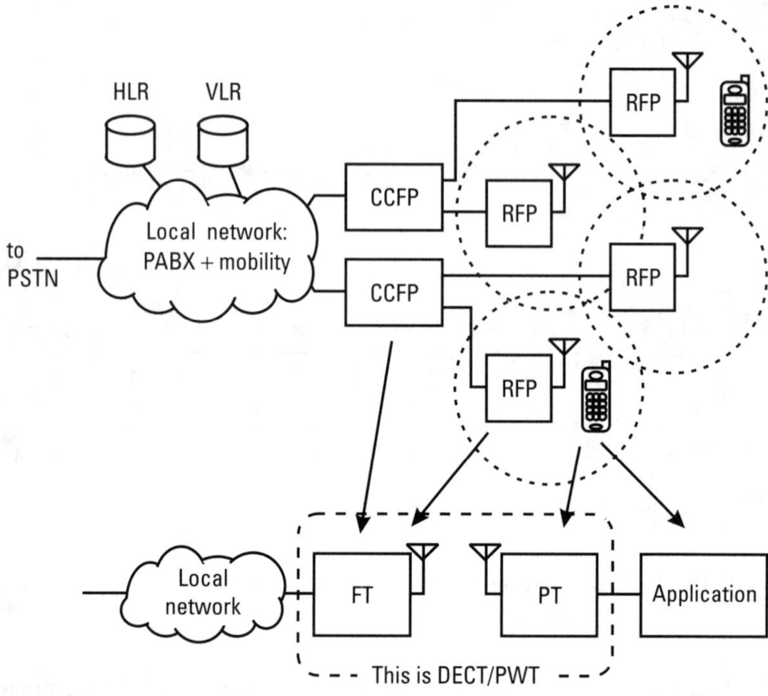

Figure 2.2 The DECT/PWT parts of a cordless PABX.

may carry power to the RFPs. The mobility management in Figure 2.2 is included as an application in the local network and is not directly specified by the standards. What is specified is a set of cordless services that support mobility management applications, such as those of a cordless PABX. In that sense, DECT and PWT are different from GSM or PCS1900. In those digital cellular systems, most parts of the local and global networks are fully specified, including global mobility management.

2.1.2 The Coexistence Requirement

The first fundamental objective of the DECT and PWT standards is to require all compliant systems to operate in such a way that independent applications can equally share the same radio spectrum. This is called *coexistence*, and the base standards [1,2] are sometimes said to provide a coexistence interface (CI). Achieving this objective is the primary function of the physical layer (PHL). Actually, the DECT and PWT standards themselves are optional and nobody is required to meet them. However, the use of radio spectrum in both the

United States and Europe is the subject of regulation. For that reason, the requirement to comply with coexistence standards is imposed from a regulatory point of view in both places. The requirement is achieved in different ways. In the United States, coexistence is achieved for PWT and for all systems that share the PCS band through separate Federal Communications Commission (FCC) regulations [3]; in Europe, it is achieved through a European Union regulation based on the DECT PHL [4].

The usual radio specification items apply here: permissible cochannel interference, adjacent channel interference level, radio sensitivity, transmitter power output, and other familiar RF (radio frequency) specifications. However, the reader of the standards soon will see that the operation of the protocol is dynamic in its use of the RF spectrum. Occupying a channel in the PHL, therefore, is the subject of certain procedures at the PHL, rather than just operational radio parameters. For that reason, procedures such as channel acquisition are specified in the PHL and other parts of the DECT and PWT standards.

Under the definition used by DECT and PWT, coexistence also requires some compliance in the next protocol layer, the medium access control (MAC) layer. The MAC layer specifies the very first level of data structuring on top of the data packets, which are the highest level data structures defined by the PHL. Complying with the lowest level of the MAC layer allows other DECT systems to recognize CI-compliant equipment as a DECT or PWT system and also to recognize that it does not comply with any other part of the standard. Why should that be considered part of coexistence? The answer is that DECT systems take note of the environment around them, in particular the operation of other similar systems. It is important that they do that to maximize the use of the radio spectrum. Some knowledge of other systems is important, so DECT and PWT rely on those other systems, including other DECT and PWT systems, having a certain structure and behavior.

2.1.3 An Opportunity for Interoperation

The second objective of the DECT system is to permit interoperation of different cordless systems from different manufacturers. Not all applications benefit from interoperation, so the base standard is selective and is written in such a way that it offers interoperation if you want it for any application but does not demand it. Other DECT and PWT standards specify interoperation for particular applications, and the base standard provides the means to achieve it.

The base standard creates the principle of *provision optional, process mandatory*. To comply with the standard, you can leave out any optional feature; however, if you do provide an optional feature, you must provide it in the specified way. What does that mean? It means that whenever a feature is provided in both ends of a cordless link, the feature is available to the user. If neither end provides an optional feature, then clearly the user will not be able to use it. That is also true if the procedure is provided at only one end of the system. This situation introduces another fundamental requirement to all DECT and PWT systems. They must ignore features they do not support in such a way that no misoperation of any other feature occurs. This requirement cannot be overstated. When a cordless system with which yours is communicating tries to invoke a feature that your system does not support, your system must ignore the attempt and go on to the next relevant operation without any misoperation.

There is also an implicit requirement here that if your system's attempts to operate a feature get no cooperation, your system must gracefully conclude that the other system does not support the feature and continue, if possible, without misoperation but without using that feature.

2.1.4 Interoperation for Certain Applications

The third objective of the DECT and PWT standards is to specify interoperation of systems in certain applications. It does that by means of separate application-specific *profiles*. Profiles are simply other standards that apply only to equipment capable of providing specific services. The most important profile is probably the GAP [5], which applies to all systems that are capable of providing normal voice telephony. The U.S. version, the CPAP, is slightly different [6].

If the equipment you are producing is not the subject of a profile, then coexistence is the minimum level of compliance demanded of your equipment. That means you must comply with the PHL and the first level of the MAC layer up to the MAC-layer *trap door* (see part 1 of [1]). If your equipment is permitted by regulation to not comply with any further parts of the standard, then the sharing of spectrum between equipment will work fairly, but there can be no interoperation of equipment. Such equipment is called *CI base compliant.*

Equipment that conforms to the base standards and to an application-specific profile standard (e.g., the GAP) is said to be *CI profile compliant.*

2.1.5 Regulation of Interoperation

In the European regulatory environment, some communications applications are considered to be fundamentally important, voice telephony being one of those applications. The European regulatory regime goes further than the optional ETSI DECT standards. Common Technical Regulations are published that require the interoperation of equipment from different manufacturers to allow users to have guaranteed access to basic telephone services. Such regulations do not exist for PWT in the United States.

2.2 The Radio Interface

So far, we have concentrated on some of the aims of the DECT and PWT systems. Now we introduce the way in which those systems use the radio spectrum. Formally, the standards are "layered" in a similar fashion to the ISO (International Standards Organization) layered protocols, and the radio interface is formally the PHL. (Chapter 3 examines layered protocols in detail, and Chapter 4 discusses the radio interface and its operation.)

The PHL specification has far-reaching consequences on equipment features, performance, and costs. The apparently simple PHL structure gives DECT and PWT several important properties:

- High capacity for office communication equipment;
- Inexpensive multichannel base stations;
- A flexible set of high-bit-rate voice and data services;
- Seamless "make-before-break" handover;
- Antenna diversity at base station end only;
- Wireless repeaters;
- Toll-quality voice.

2.2.1 Introduction to the Radio Interface

The entire purpose of the radio interface, the PHL, is to divide a certain portion of the radio spectrum into channels that can be assigned in various combinations to various cordless applications. The way in which PHL channels are assigned to applications is specified in the higher layers of the protocol. That is done to allow the PHL to provide a flexible foundation on which many different applications can be built. Hence, at the lowest layers of the protocol, very little relates to applications and the standard looks rather abstract. This introduction illustrates the PHL from the point of view of a normal voice telephony application, but remember that other applications use the PHL in different ways. A rationale behind some of the choices made in the design of the PHL are discussed later (Sections 2.2.2 through 2.2.9). Here, we cover the basic ideas of what the PHL is.

The DECT PHL occupies a block of radio spectrum from 1880 MHz to 1900 MHz (this is the European spectrum allocation; the U.S. allocation is different, as explained in Section 2.4). In Europe, 10 radio channels are allocated within this spectrum, each 1.728 MHz wide and capable of handling 12 simultaneous two-way telephone conversations, as shown in Figure 2.3(a). That is done by calling each successive 10-ms time interval a *frame*, dividing the frame into 24 regular timeslots, transmitting data within the timeslots at 1,152 kbps, and allocating two timeslots per telephone call, one timeslot to each communication direction, as shown in Figure 2.3(b). Each conversation occupies a duplex channel that carries a two-way digital voice channel, a two-way digital signaling channel, and a two-way synchronization channel. That way, up to 120 digital telephone conversations can be carried in the European DECT spectrum.

The division of the radio spectrum into 10 channels appears to make DECT an FDMA system. That, however, is not the reality of it at all. DECT

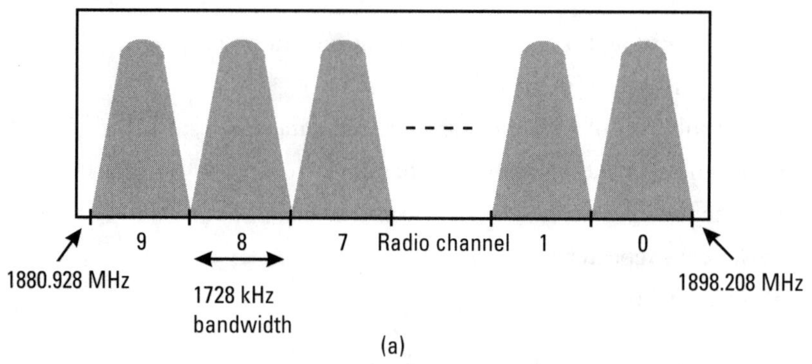

(a)

Data packets on the same RF carrier

Transmission from handset

Transmission from base

0 5 10 Time (ms) 15

(b)

Figure 2.3 Division of the DECT PHL (a) in frequency, and (b) in time.

and PWT are *not* primarily FDMA systems, even though they employ FDMA. The division of each radio channel to allow it to carry 12 telephone conversations is much more important in determining the characteristics of the system. The principal characteristic of the system, therefore, is that of a TDMA system using TDD to divide the outgoing and incoming speech paths.

Think of the repeating 10-ms frame period as a circle divided into 24 sectors, numbered 0 to 23. Then imagine stacking 10 of those circles on top of each other, one for each carrier frequency defined in the DECT system. Picturing the DECT PHL that way conveniently organizes the 240 simplex channels into a cylinder (Figure 2.4). Each horizontal slice represents one of the 10 DECT radio carriers. The circumference of a slice represents the 10-ms TDMA frame imposed on that carrier. Each sector of the frame represents a timeslot, one of the 24 imposed on the 10-ms TDMA frame. A similar arrangement can be made for PWT.

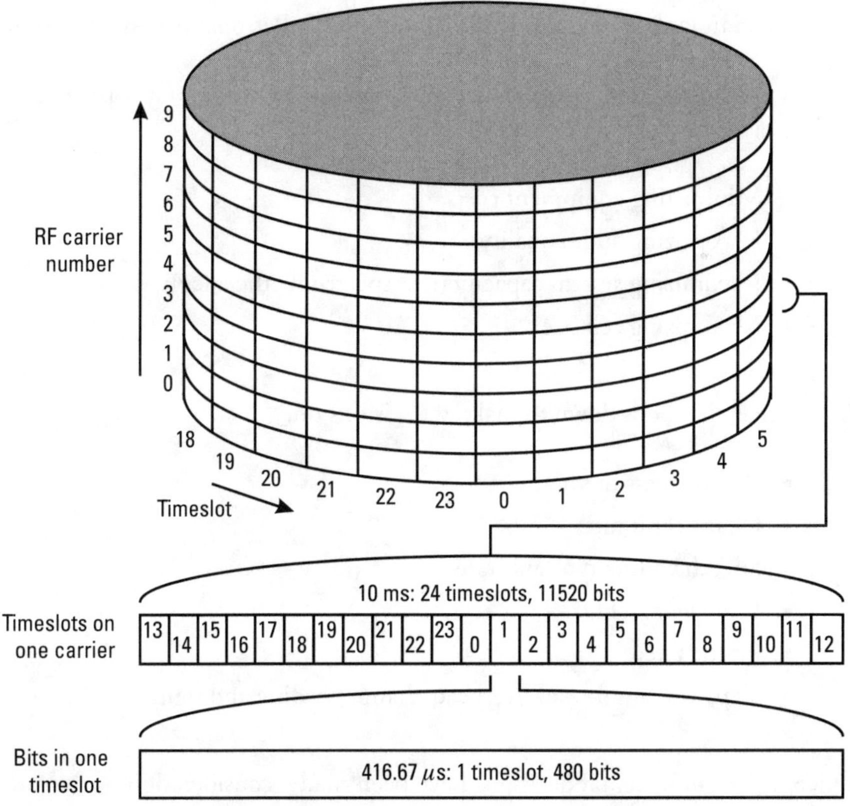

Figure 2.4 A convenient organization of the DECT physical layer channels.

The key thing to note about such an arrangement of the radio spectrum is that a single radio transceiver may carry 12 conversations simultaneously. Conceptually, it is a more complex system than one in which 12 separate radio transceivers are used for 12 different conversations. The radio transceiver has to be capable of switching from handling one conversation during one timeslot to handling another during the next if it is to be able to handle all 12. However, it ultimately is cheaper to use just one radio transceiver to handle multiple telephone links.

2.2.2 Outline of the Basic Rationale

The strengths and weakness of the DECT PHL can be appreciated by recalling the original goals and constraints of the DECT project. A discussion of why particular approaches and parameters were chosen also helps clarify the differences between DECT and other cordless standards such as CT2, the Japanese Personal Handyphone System (PHS) and the U.S. Personal Access Communication System (PACS) .

The chosen PHL design is a complex trade-off among (at least) the following factors:

- Minimizing equipment costs;
- Maximizing audio quality;
- Maximizing system capacity (i.e., spectrum efficiency);
- Maximizing cell size.

In practice, that boils down to making decisions on:

- Radio access method;
- Channel bit rate;
- Speech coding type and rate;
- FDD versus TDD;
- Frame rate;
- Type and amount of overhead channel coding and signaling.

Once those fundamental decisions have been made, consideration of the radio range required and other aspects of the radio environment (including the regulatory situation) leads to decisions on the following:

- Modulation scheme;
- Transmit power level.

2.2.3 Radio Access Method

One of the primary initial applications for DECT was to provide wireless office equipment, like PABXs and local area networks (LANs). The goal was to provide office equipment that offered, as far as possible, a no-compromise alternative to a wired phone, and products had to be able to offer audio quality, capacity, and security practically indistinguishable from those offered by wired extensions. Such wireless PABXs were anticipated to carry large amounts of traffic, so one of the key requirements was to create a relatively inexpensive multichannel base station. In a residential system, where it is likely that no more than a single user uses the telephone at any time, a single-channel base station is adequate. In an office system, on the other hand, it is likely that several users within range of each individual cell will want to make and receive calls.

The problem thus became one of designing the most cost-effective multichannel base station. One of the most expensive components of a base station channel is the radio transceiver, a piece of precision analog engineering. The other sections of the base station (baseband and telephone line interface sections), although in some ways much more complex than the RF section, benefit enormously from developments and cost reductions in similar products (such as personal computers and wired telephones). The likely cost of the transceiver suggested the use of a TDMA approach, since that would allow several users to share one radio transceiver and several users would share the cost. FDMA could not do that. Using TDMA not only produced a major cost-saving for large numbers of channels but led to a number of other important advantages.

CDMA was not a serious contender, since at the time it was considered too complex to provide the low-cost products required. In any event, DECT and PWT, having chosen TDMA, still provide many of the benefits that are now claimed for CDMA systems (excellent frequency reuse, no fixed-frequency planning, seamless handover, and high capacity).

2.2.4 Channel Bit Rate

While the choice of TDMA over FDMA and CDMA (Code-Division Multiple Access) was a fairly obvious choice, the choice of the exact number of channels was a much more delicate balance of many important issues. On the face of it, one might want to increase the number of TDMA channels as much as possible: the greater the number of TDMA slots per carrier, the greater the number of users who share a single transceiver and its cost. If you need to support many users per transceiver, a system with 30 TDMA channels per carrier is more cost effective than a system with 3 TDMA channels per carrier.

However, increasing the number of channels gives rise to some practical problems. First, a large number of channels per carrier means a high data rate, and the faster the transmitter bit rate, the more distortion in the received signal due to the radio channel. Second, the more users sharing a single carrier, the greater the delay before each user is given a turn, so the greater the transmission delay. Moreover, as the receiver bandwidth is increased to accommodate a wider band signal, the level of the thermal noise floor is increased, which reduces sensitivity. As the bit rate is increased, transmitter power then must be increased to keep the energy per bit constant. Also, because more of the division is done by time rather than by frequency, it is important that neighboring systems are synchronized to prevent mutual interference.

Increasing the number of channels by increasing the aggregate bit rate and reducing the duration of each individual bit period gives rise to problems with time dispersion of taken too far. As the bit length gets shorter, it becomes comparable with the length of time needed for the signal to propagate from transmitter to receiver. Since the radio signal propagates via various paths of different lengths, the receiver may receive several "echoes" of the same signal. Time dispersion starts to become a problem in cases where the cell size means that the potential time dispersion delays can be comparable with the bit period.

For example, if the difference in path length between two radio propagation paths is 600 m, then at 3×10^8 m/s the differential delay is 2 μs. If the bit period is 1 μs, then the receiver may receive the same data twice, one copy delayed 2 bits after the other and the second interfering with the first. Equivalently, the receiver also would experience a 2-bit delay if the bit rate was increased to give a bit period of 0.1 μs and the difference in path length was 60 m. That effect is common in digital macrocellular systems, where a time dispersion "equalizer" is used to compensate for the effect. DECT's designers decided not to increase the bit rate to the point where an equalizer would be necessary for typical applications (such as residential and office systems) but left the flexibility whereby it would be possible for manufacturers to choose to add it to their products to give extended range performance if needed. An aggregate bit rate of 1,152 kbps was settled on. With the maximum bit rate having been decided, the number of channels is a function of the bit rate for each individual user channel.

2.2.5 Speech Coding Type and Rate

The goal of providing a no-compromise alternative to the wired office extension called for high-quality voice at a reasonable cost and with reasonable spectrum efficiency. That led without much difficulty to the selection of 32-kbps ADPCM for the speech coding. It is a relatively common and inexpensive technique. This voice codec provides very high quality voice that most people can hardly distinguish from wireline quality. It also allows the passage of voice band signals such as DTMF tone signaling (to access voice mail, etc.) and some low-rate voice band modem and fax signals. ADPCM, an enhancement of the ubiquitous standard 64-kbps pulse code modulation (PCM) used in many of the world's telephone networks, uses a compression algorithm to reduce the channel rate from 64 kbps to 32 kbps. ADPCM is one of the classes of waveform coders that predicts the next signal value, samples the actual signal, and transmits the difference. Waveform coders are relatively simple and have a low processing delay between sampling the analog input and producing the digital output.

It would have been possible to select a more spectrum-efficient speech coding, such as an 8 kbps or 16 kbps codec. Many of the low-bit-rate codecs are programmed to recognize various characteristics of human speech; by eliminating redundant information, they minimize the bit rate. However, such complex coders also have a significant delay between sampling the analog input and producing the digital output and consume a lot of power. They commonly are used in digital cellular systems, where spectrum efficiency is at a premium. It was felt that DECT, because of its use of small cells, would achieve high levels of spectrum efficiency and that the high additional cost of an expensive vocoder would be more than the market for residential cordless phones could bear. It also would have been possible to choose 16-kbps delta modulation; however, the audio quality was not as good, and implementations were not as supported by component manufacturers.

The standard nevertheless makes provision for a "half-rate" slot (whose capacity is actually is smaller than half that of a 32-kbps full slot), which could carry a low-rate speech codec if required. The standard also allows for a double slot, which allows conventional PCM to be carried at its full rate of 64 kbps. That makes it possible to use DECT to provide fully transparent wireless replacements for wired telephone connections. In other words, it allows DECT and PWT to provide a wireless "socket" into which you can plug any conventional piece of standard telephone equipment (fax, modem, etc.) in the full confidence that it will work as expected.

2.2.6 Duplexing

Fixing the total transmission rate and the rate for each individual channel more or less fixes the number of channels that can be carried. First, however, it is necessary to decide whether the available transmission bandwidth is to be used for one direction or for both directions. If the maximum channel rate is 1,152 kbps per carrier, it must be decided whether a single carrier should be used for transmissions from base to handset and a second carrier for transmissions from handset to base, or whether a single carrier should be used for both directions of transmission. Thus, it is necessary to choose between FDD and TDD. In practice, the DECT system chose to use TDD.

Duplexing is the process of providing two independent channels, one for the downlink channel (i.e., the path between the transmitter in the base station and the receiver in the handset) and one for the uplink channel (i.e., the path between the transmitter in the handset and the receiver in the base station). In FDD, separate radio frequency carriers are used for the uplink and the downlink. That is the traditional method of providing a duplex channel in cellular radio systems. The main disadvantage with that approach is the need to use a duplex filter in the handset. The duplex filter is normally a costly and bulky component that is there to separate the transmitter chain from the receiver chain. Another characteristic of FDD is the fact that the distortion experienced by the downlink is different from that experienced on the uplink (Figure 2.5); in a TDD system, that effect is minimized (although not eliminated).

In TDD, a single channel is used alternately for both uplink and downlink. That setup is similar to old-fashioned simplex radios, in which one user would key the mike, then say a message terminated with the word "over" to invite the other party to answer. In TDD, to give the illusion that both users are using the channel at the same time, the digital speech has to be transmitted twice as fast, to give enough time for the reverse channel to also use the carrier. Using the greater channel bit rate may lead to time dispersion problems. Doubling the channel bit rate also doubles the required channel bandwidth, and that doubles the noise floor. All other things being equal, it reduces the radio receiver sensitivity by 3 dB, which reduces maximum cell size.

One key advantage of TDD is the fact that since the same radio channel is used in both directions, distortion on the uplink is similar to the distortion on the downlink. The distortion is not identical, however, because in the time period between the uplink transmit burst and the downlink transmit bursts the channel conditions may have changed slightly. For the low-speed mobility applications with which DECT is intended to cope, the change in channel conditions is small enough. The fact that the channel is nearly symmetrical makes it possible to implement channel equalization techniques at only one end of the

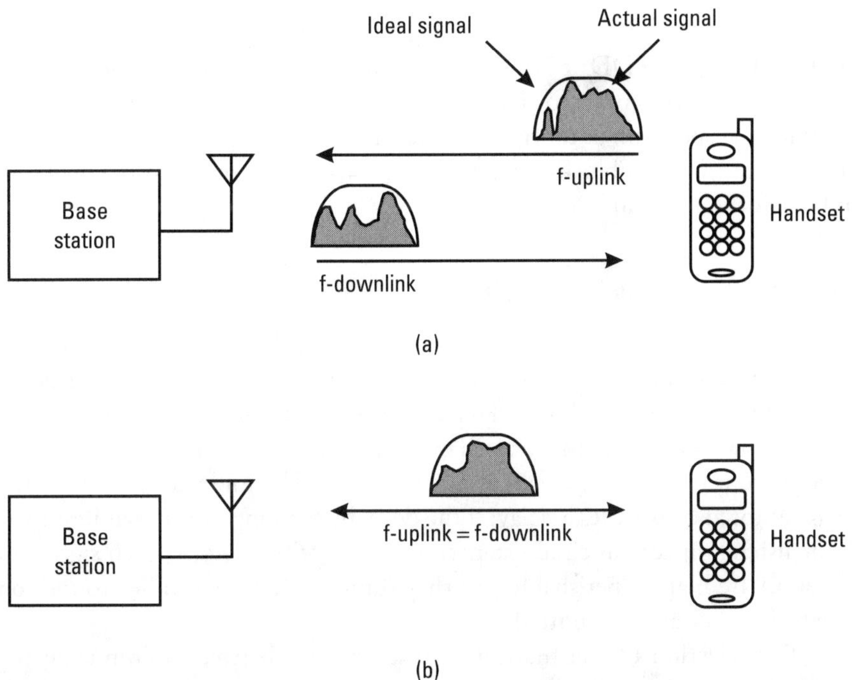

Figure 2.5 (a) Frequency division duplexing versus (b) time division duplexing.

link. That can be important because it allows more complex and expensive techniques to be implemented only in the base station, thus allowing the cost of the handset to be reduced. If a base station can serve a total population of 50 handsets (although not all at once), the cost of any features implemented in the base station is shared among the 50 users.

It is possible, for example, to implement antenna diversity at the base station end only. Antenna diversity is a simple but effective technique to combat signal fading (described in Section 8.3.1). If a base station with two independent antennas selects the one from which it receives the stronger signal, then that also should be the antenna from which the handset receives the stronger signal. Antenna diversity is possible at the handset but difficult to implement since the handset normally is too small to allow the antenna to be separated by the minimum quarter-wavelength separation needed for reasonable performance. The symmetrical channel also offers the possibility of implementing a channel predistortion equalizer at the base station. The equalizer measures the distortion on the uplink, calculates the inverse of the distortion, and applies that inversion to the transmitted signal so the signal is received without distortion at the handset.

Another practical advantage of TDD that encouraged its final adoption is the fact that with TDD it is necessary to identify and allocate only a single block of radio frequencies for the system rather than a duplex pair. It is much easier for the spectrum authorities to find or release a single block of spectrum than it is to find two blocks separated by a particular spacing (one for uplink and one for the downlink).

2.2.7 Frame Rate and Packet Size

One of the final issues to resolve in the decision on the number of channels to use and whether to use TDD or FDD is the frame rate. The frame rate is the rate at which each user is allocated regular slots to transmit and receive information. The frame rate fixes the interval over which the transmitter buffers (stores) data before it transmits the data as a data packet. The interval over which the data are stored constitutes a delay; if the delay be too long, it can manifest itself to the listener in certain circumstances as a perceptible echo. Speech signals in any event are highly "perishable," so they cannot be allowed to lie around too long before they are transmitted.

The selection of the frame rate is a trade-off between minimizing the potential production of undesirable echoes and maximizing the efficiency with which the channel is used. If we need to use a particular overhead per data packet for signaling and synchronization, then efficiency dictates that we must wait for enough payload data to arrive to keep to a minimum the relative amount needed for the overhead. Figure 2.6 illustrates this trade-off.

Reducing the frame length reduces the period of time over which speech data are stored before transmission and thus minimizes delay and any echoes that may occur. Increasing the frame length maximizes efficiency.

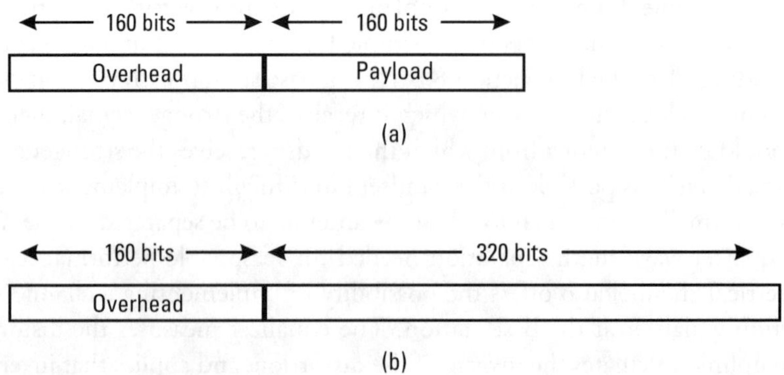

Figure 2.6 (a) Packet with small payload versus (b) packet with large payload.

The actual overhead data comprise a header field plus a guard interval between successive transmission bursts. For reasons that will be explained later, the overhead information is, as we have assumed, a fixed length and not a proportion of the user data burst. For example, if 160 bits of overhead data were necessary, then if the user data payload were 160 bits long, the ratio of wanted payload to total data would be only 50%. Doubling the length of the payload data to 320 bits increases that efficiency ratio to 67%. The figure of 320 bits is the actual payload size used by DECT and PWT for speech traffic.

Increasing the length of the payload section increases the frame length, the time between a user's regular slots and thus the delay seen by the user's data. If the frame length is considered to be too long, then it is necessary to reduce the number of channels in the frame. For example, if each user is allocated a regular timeslot of say 0.417 ms (a value that is actually used) and if we were then to have 12 duplex channels (again, an actual number that is used), then using TDD we have 24 timeslots (12 for the downlink and 12 for the uplink), giving a total frame length and thus a speech buffering delay of 10 ms. Such a delay is considered rather long for most speech connections, but it is acceptable with the correct degree of echo control. If we had chosen to use, say, 24 duplex channels instead, the buffering delay would have risen to 20 ms, which, although possible, would have increased the complexity of the echo control required.

2.2.8 Channel Coding and Signaling

Once the frame structure has been decided, it is necessary to decide the structure of individual channels, especially the signaling and control channels. One of the key decisions made on the DECT interface was to preface each transmission of user data with a header of synchronization and signaling data, which serves two purposes:

- To allow the handset and base station to regularly synchronize to each other;
- To transfer control information.

The header is overhead information, a necessary evil that occupies bandwidth that could otherwise be occupied by user traffic. It would have been possible to operate without synchronization within the header by establishing a connection between handset and base station and then relying on both ends to maintain independent free-running clocks synchronized only to the user traffic; indeed, that approach is used in other similar systems [7]. Operating without a synchronization header would increase spectrum efficiency by increasing the number of channels. That approach, however, is vulnerable to the problems caused by the fragile radio channel. If the radio signal fades and both ends temporarily lose contact and slip out of mutual synchronism, it becomes difficult to reestablish synchronization without going back to the beginning and setting up the link all over again.

Any such failures of synchronization and delays in the processes needed to reestablish the connection can have a major impact on users' subjective perception of the quality of the radio connection. There is, in fact, a major subjective difference in the way subscribers perceive the effect of a poor radio channel depending on whether it is digital or analog. The analog channel experiences a gradual worsening with the level of background noise rising and masking the conversation. As bad as that is, users at least understand that the signal is too weak and that they should move into a better coverage area. Digital transmissions tend to have a binary nature: they are received either perfectly or not at all. Furthermore, a digital transmission is a stream of regular and separate packets of data. If one of the packets is lost, the user experiences a conversation punctuated by a "dropout." Most users consider dropouts incomprehensible, frustrating, and irritating. That frustration is exacerbated if just one channel from the uplink and downlink is affected; one party continues talking in the belief that the other can still hear. Many users find intermittent interruption much more difficult to cope with than a rise in the level of background hiss.

DECT and PWT provide mechanisms to avoid dropouts, and mechanisms to minimize the duration of any interruptions that do occur. To prevent dropouts from occurring in the first place, DECT provides the *extended preamble mechanism.* Explained in more detail in Section 4.5.4, extended preamble mechanism basically means that the payload data are prefaced with a pair of identical header sections (preambles). The receiver uses a separate antenna to receive and measure the quality of the signal in each preamble and chooses the better of the two antennas to receive the rest of the packet. This is a powerful method because it allows the receiver to avoid fades in the radio signal. The use of a separate synchronization field after the preambles thus allows both ends to stay tightly locked together.

The presence of the preambles and the synchronization field in each packet allow each single packet to be received and demodulated correctly, which allows fast call setup and handover. The mechanism has proved in practice to be an effective method of maintaining a stable radio channel and one not punctuated by dropouts. These techniques contribute greatly to what users consider as wireline speech quality.

The signaling field in each packet is used to transmit two things. The first is a set of common system information for all users to see. This is called the *beacon* and is transmitted on all channels in use rather than on just one or more dedicated signaling channels. The second is the signaling for a particular call of which the packet is a part. The duplication of common system information avoids a single point of failure and makes DECT and PWT extremely resistant to interference on the common information channels. A system that relies on a single channel for system information is extremely vulnerable if the channel is accidentally or maliciously interfered with. While most macrocellular systems do use dedicated common information channels, such networks normally are carefully planned and managed by a single network operator and are not subject to interference from uncoordinated users. However, the DECT/PWT system allows public operators to share spectrum with uncoordinated private users with the confidence that the system will not be degraded.

For both the signaling and user data, there is no complex channel coding to protect against errors. It is assumed that the nature of the radio channel is that either a packet will be received without errors or it will contain too many errors to correct. To protect the signal against so many errors, it would be necessary to split the signal and transmit the parts over a series of packets. However, that not only would require complex coding circuitry but would increase the delay between an analog voice signal being digitized at one end and being completely reassembled and reproduced at the other. That extra delay would be perceptible as an acoustic delay and would require more sophisticated echo controllers.

Between data packets, there has to be some guard space. The guard space of approximately 50 μs was chosen as a trade-off between spectrum efficiency, cell size, and product cost. It would have been possible to reduce the guard space by introducing timing advance mechanisms, but that was considered too complex for a system intended for low-cost mass market use. Subsequently, an optional timing advance mechanism has been introduced to deal with advanced applications.

2.2.9 Dynamic Channel Allocation

DECT and PWT are intended for use in uncoordinated systems that might be neighbors or that might overlap significantly in their radio coverage. They use dynamic channel allocation (DCA) rather than fixed frequency planning. Before a radio channel on which to transmit is chosen, all the radio channels potentially are available to all equipment. The equipment makes the choice, which depends on the signal strengths in the local radio environment.

The TDMA/TDD structure makes possible a "continuous" monitoring of the quality of all potential radio channels in the intervals between the transmit and the receive timeslots. During unused timeslots, the receiver is idle; having nothing else to do, it can monitor other radio channels. For example, a handset, while engaged in a conversation, can find out if any alternative radio channels are available to its current base station that are clearer of interference. It can then initiate a channel handover to improve service quality. It also can monitor nearby base stations, looking for one that could offer a better link quality. On finding a better candidate, the handset can initiate a handover and take service from the new base station. DCA is therefore at the heart of handover. Both channel allocation and handover (base station allocation) are driven by the handset, which continuously monitors the alternative radio channels and alternative base stations in its neighborhood. Thus, continuous DCA provides excellent spectrum usage without fixed frequency planning.

Not only does DCA simplify the process of frequency planning and coordination, it gives important performance advantages. The most obvious advantage is that the best channel continuously is selected, which gives excellent voice quality. Second, it can increase system capacity, which can be explained as follows. At the limit of range, the desired signal power received is just enough to overcome the detrimental effects of thermal noise and interference signals. That is, the actual value of carrier-to-interference ratio (C/I) is equal to the minimum required value. As the portable comes closer to the base station, it has more C/I than it really requires. The closer the portable is to the base station, the more interference it can withstand. The system's continuous dynamic channel measurement means that the portable is constantly choosing a channel with the best C/I, providing maximum protection from interference and thus maximizing system capacity. The DCA is automatically exploiting the excess C/I close to a base station to increase system capacity.

Older fixed channel assignment schemes mostly do not exploit that extra power. Power-controlled systems, such as IS-54 or GSM, simply reduce the transmitted power when the portable is closer to the base station, reducing the total amount of interference generated and thus increasing potential system capacity. DECT and PWT do that as well, since they both have a low-power mode if they are close enough to a base station.

As a result of the complete DCA scheme, DECT and PWT share many of the advantages claimed for CDMA systems. They require no frequency planning, and they provide high capacity through intelligent selection of the optimum channel.

2.3 Basic System Operation

You do not have to know the details of the protocol itself to understand how, at the top level, a DECT or PWT system works. The full details are presented later on, in Chapters 3 to 7. Here, however, is a description of the basic principles.

2.3.1 The Fixed-Part Beacon

Every radio cell, from a single-cell residential base station up to each cell of a multicell system, broadcasts a beacon all the time it is powered up and operating. The beacon is an essential part of mobility management and serves several purposes.

First, if the portable can see the beacons of many cells, it can determine which has the strongest signal and therefore decide which is the best cell on which to lock. The next thing is that the portable can determine from decoding the beacon's signals the identity of the location area in which it currently resides. The location area levels (LALs) that the beacon broadcasts are assigned when a system is installed and perhaps occasionally reassigned during operation, especially if the system is extended or otherwise reconfigured.

The beacon also broadcasts a system identity (sometimes it broadcasts several identities) called an access rights identifier (ARI). Like the LALs, the system's ARIs are assigned when the system is installed and possibly updated occasionally during operation. Before it can be used to make and receive calls, a portable has to store the identities of all systems it is allowed to access; thus, the portable knows whether a system will accept its connections. As part of the subscription process in which the portable is given the identity of a system that it will be allowed to use, the portable's own unique identity is stored in the system's database of subscribed portables so the system too can determine if the portable is allowed to use it.

The beacon also contains information about how the fixed system operates. It carries information about which optional services the system supports, details about how its radio receivers scan the radio channels for outgoing calls, how many radio receivers are currently free, and much more. Those details allow a portable to predict on which of the timeslots and on which of the radio carriers a system will be listening at any one time, which allows the portable to rapidly request that a radio link be set up.

Finally, the beacon carries paging messages for portables, and the portable knows from the beacon's structure precisely at which times any paging messages will appear. The most important use of these paging messages is to tell a portable that an incoming call is available and the portable should set up a radio link to alert the user. Knowledge of the beacon's timing details allows the portable to synchronize its sleep/wake cycle so that it spends most of the time saving power and looks for paging messages only when there is a chance of a paging message being sent. By sending paging messages only at certain intervals, the fixed system allows portables to greatly reduce their power consumption and extend their battery life.

In general, then, the beacon acts as an invaluable source of information to tell the portable where it is, how best to access systems for which it has been given access rights, and how to save power by turning off at times when the fixed system will not be broadcasting paging messages.

2.3.2 Portable Lock-On and Attach at Power-Up

Looking at matters from the portable's point of view, the aim of every portable that has just powered up is to find a home. It has to identify a good beacon, gain and retain timing synchronization with the system, and inform the system that it is present in a particular location, ready to accept incoming calls. That is done with the aid of the data on the fixed part's beacon. It is hard to underestimate the importance of the system's beacon and the information it broadcasts in influencing the operation of the population of portables it serves.

A newly activated portable will scan all the available timeslots on all the RF channels to see if there is a beacon whose broadcast identities it recognizes as one of the identities it currently stores. Furthermore, it will look for the recognized beacon with the highest RF power level, which is assumed to be from the closest cell and so will give the most reliable service. For simple residential telephone sets with a single cell centered on the only base station, there may be no option. Nevertheless, the handset cannot make the assumption that it is operating in a single-cell environment, because it soon may be used to make and receive calls in a multicell system to which it also has a subscription.

Once the handset has selected a suitable beacon, it locks its receiver scanning behavior to the beacon and transmits a message to the fixed part to attach to the system. Attaching informs the system of where the portable is, its identity, and the fact that it is ready to receive incoming calls.

Now that its timing is locked to the fixed system's beacon, the portable can spend much of its time asleep and not consuming much power, waking up only at intervals to see if there are any paging messages. The fixed system normally imposes a long cycle (160 ms) on the transmissions from its beacon, called the multiframe. Only at one specified point in the multiframe cycle will the fixed system begin broadcasting paging messages indicating that an incoming call is present. That is the only point at which the portable has to wake up during the superframe cycle. If there are no paging messages, it goes back to sleep again and saves power. Occasionally, the portable reverifies the complete system information by staying awake during the entire multiframe cycle, and occasionally it may search for a better beacon and attach to it instead.

If an attached portable goes out of range of its current system at any time, it will attempt to lock on to and attach to any other system it recognizes. If it cannot recognize a system, it keeps waking up at intervals to see if it has moved back into range.

2.3.3 Making an Outgoing Call From a Handset

Making an outgoing call, that is, one originating from the portable, requires the portable to be locked to the fixed system, to know that a free radio receiver will be scanning for calls, and to identify its scanning pattern. All that is known to a portable that is attached to a system.

When the portable's user presses the line button, the portable selects a free radio channel on which to make a call setup attempt, and the portable waits until it knows the fixed system is listening to the selected channel.

First, the portable (actually, its MAC layer) sets up a bearer for the requested service. For a voice connection, usually that has a duplex 32-kbps user channel with an attached 6.4-kbps control channel, which fits into a single pair of slots within the 10-ms TDMA frame. Once the bearer is set up, the portable's data link control (DLC) layer sets up an automatic repeat request (ARQ) protection scheme for signaling messages within the control channel, to make sure they are not lost or corrupted.

The portable and fixed network (NWK) layers then exchange all the information needed to set up the call, by exchanging NWK-layer messages within the protected control channel and retransmitting any that are corrupted or otherwise lost. The fixed part uses that information to connect the call. When the call is connected and there is a useful end-to-end connection, the NWK layer instructs the portable to connect audio to the user via the user channel.

2.3.4 Incoming Call From the Fixed System

There really is no such thing as an incoming call setup for normal voice calls. For that type of service, all calls ultimately are originated by the handset, which is notified of the incoming call and asked if it wants to accept it. That is done by a process in which the handset is paged.

As we have already seen, a logical channel is attached to each beacon that is dedicated to the broadcast of paging messages for handsets. The most frequent use of that channel is for delivering incoming call requests. The same paging information usually is delivered simultaneously on the beacon channel over many or all radio cells in a system, so a handset attached to any cell in the system will see it. In really big systems, paging capacity is preserved by sending the paging messages for a handset over only a part of the fixed system, that part where the system knows the handset to be. In small systems, it is simpler to send paging messages out from every cell.

Once a handset alerts the user and the user takes the call by pressing the line button, the call setup procedure is precisely the same as in an outgoing call. The only difference is that the fixed system has an incoming call to connect instead.

2.3.5 Mobility Management and Handover

The principle of DECT/PWT's mobility management is that the portable is in charge of all things to do with the management of its own mobility. That does not mean that the fixed part has no important role, just that the fixed part does not take steps to locate the portable or to determine if it needs action with respect to mobility management. The portable always knows when it has made a significant change in location, since the best beacon it can see will broadcast a new LAL compared to the one it last saw. That prompts the portable to reattach to the system and so inform it that it has moved location.

The continuous search for a better beacon continues even when a call is in progress. Whenever a locked and attached portable participates in a normal voice call, it does not need to be active for most of the 10 ms of the TDMA frame. It needs to transmit on only one of the 24 timeslots and receive on another. The free time it has can be devoted to searching for an alternative beacon from the same system and assessing whether it may offer better communication quality by virtue of a better signal strength. If a better beacon can be seen, the portable may wish to initiate a handover to transfer the communication to the new cell.

Note that in DECT the portable, not the fixed part, makes the decisions on handover. That is simply to distribute the processing power required to make the decision around the portables rather than concentrating it in the fixed part. Such a system should be able to handle a larger number of portables with a given system processing power.

2.4 The Differences Between DECT and PWT

There are, in fact, three personal wireless communications systems in the DECT class: DECT, PWT and PWT-E. PWT-E is also known as DCT1900 (for Digital Cordless Telecommunications at 1900 MHz).

PWT and PWT-E are the systems used in the United States. PWT-E is used by network operators who have purchased exclusive licenses to a block of radio spectrum to provide telecommunication services. PWT is the standard used in the spectrum band which is shared among all potential users rather than being licensed exclusively to a single organization. This distinction gives rise to the term *unlicensed* to describe PWT and *licensed* to describe PWT-E. There is a set of FCC regulatory rules (called an etiquette) [3] for how to share the unlicensed spectrum among all applications using it, including PWT. The etiquette has two variants covering two separate subbands, the *isochronous* subband and the *asynchronous* subband. Devices that transmit at regular intervals, such as time-division voice systems, are classified as isochronous, while devices that transmit irregularly, such as LAN data systems, are classified as asynchronous. The PWT system already has its own spectrum-sharing rules within the standard, but the FCC regulations have to be followed as well. The etiquette rules do not apply to the PWT-E.

DECT, the European version of the standard, is used in all environments and for all applications. There is no division of the radio frequency depending on its use, and the only rules for using the spectrum are those contained within the standard itself.

The differences among the three versions, DECT, PWT, and PWT-E, principally are in the PHL. The changes for PWT have the effect of increasing cell size and system capacity. Of course, there also are differences to account for the different frequency bands used. The main differences are in PWT's use of a more complex and more spectrum-efficient modulation techniques and in PWT-E's use of higher transmitter powers. Other differences are mostly at the MAC layer of the PWT protocol, which operates the PHL. The differences have to do with the designation of the different radio channels in the control signaling between the fixed and portable ends and in the different rules imposed for selecting and using the radio spectrum.

2.4.1 Background

Three different versions of the standard have arisen first because of market pressures and the opportunity to develop metropolitan personal communication systems, and second because of the continuing pressure to increase the efficient use of radio spectrum (i.e., squeezing more users into the same bandwidth). The variations also arise from differences in the radio spectrum regulatory environment between the United States and Europe.

DECT was the first version to be developed. At the time it was being developed, European law had already reserved a relatively generous allocation of 20 MHz designated for its use. Although it was intended to serve several market applications (residential, office, public access, and wireline replacement), its development was driven principally by the requirements of the wireless PABX. Very high traffic loads were anticipated for office systems, and it was judged that the traffic could be served by using small cell systems rather than by attempting to select a more efficient modulation scheme. In that way, it would be possible to allow high capacity in office systems without increasing the product cost of simple residential-type products. The designers were reluctant to add increases in complexity, which could increase product cost to the point where products would not sell in large volumes. By and large, they chose simple tried and tested components and techniques. The standards-making process thus concentrated on providing service flexibility rather than on very high radio channel efficiency.

The U.S. versions were developed later, when both market pull and technology push had intensified. There was much more pressure on spectrum and, fortunately, much more experience with more efficient modulation schemes. The spectrum regulatory regime also was quite different in the United States from that in Europe. The European authorities had accepted the market for cordless products and had cleared a band to be used exclusively by DECT products. By the time PWT was proposed as a standard, the United States had already taken a much more market-oriented approach and had divided the personal wireless spectrum thus:

- The 1910 to 1930 MHz band, to be shared by all equipment that obeys the regulatory rules for spectrum sharing (the unlicensed bands);
- The 1850 to 1910 MHz band plus the 1930 to 1990 MHz band, licensed exclusively to a single organization that would use them to provide telecommunications services for a fee (the licensed bands).

In the United States, licensed bands are auctioned to the highest bidder. The prospective PCS operator purchases a license for a block of spectrum and is free to use whichever technology is best suited to its needs. The PWT-E standard was developed to suit the needs of prospective public network operators. Its alternative name, DCT1900, no doubt was chosen to be reminiscent of its big brother, PCS1900, the digital cellular system. The name also emphasizes the fact that ETSI was already working on standards to allow the integration of cordless and cellular systems. This integration takes the form of dual-mode handsets to allow one handset to access both types of network, plus the signaling standards to allow a handset to notify a cellular network that it is available to receive incoming calls in a cordless system. The cellular system then can divert an incoming call to the cordless system, and the user will take the call with a cordless phone rather than a cellular phone.

2.4.2 The Radio Interface

The principal change made to DECT for the PWT and PWT-E bands was in the modulation scheme which was changed from GMSK (Gaussian minimum shift keying) to $\pi/4$ DQPSK (Differential quadrature phase-shift keying). The $\pi/4$ DQPSK scheme is a more efficient form of modulation that permits the same amount of information to be transmitted in a narrower channel. The change from a two-level modulation to a four-level modulation halves the symbol rate on the air interface to 576 ksymbol/s, while keeping the data rate at 1,152 kbps. A comparison of the basic parameters of the DECT and PWT physical layers is given in Table 2.1.

Table 2.1
Comparison of DECT and PWT

System:	DECT	PWT	PWT	PWT-E
Etiquette	No	Asynchronous	Isochronous	No
License	No	No	No	Yes
Frequency band (MHz)	1880–1900	1910–1920	1920–1930	1850–1910, 1930–1990
No. of carriers	10	8	8	24 bands of 5
Channel spacing	1.728 MHz	1.250 MHz	1.250 MHz	1.000 MHz
Modulation	GMSK (BT = 0.5)	$\pi/4$ DQPSK ($\alpha = 0.5$)	$\pi/4$ DQPSK ($\alpha = 0.5$)	$\pi/4$ DQPSK ($\alpha = 0.5$)
Handset frequency accuracy	±50 kHz	±18 kHz	±18 kHz	±18 kHz
Standard peak power	250 mW	90 mW	90 mW	90 mW

While complying with its own standards, unlicensed PWT also must comply with part 15 of the FCC rules [3], which introduces slightly different requirements for operation in the isochronous and asynchronous bands. The fundamental rules for PWT are those of the isochronous band. Extension of PWT operation into the asynchronous band is permitted if the asynchronous band's rules are also obeyed, but the authors know of no devices that take advantage of that.

No specific details of how to comply with part 15 of the FCC rules are provided in the PWT standard. However, the spectrum-sharing etiquette requires, among other things, the use of the extended radio channel plan and the ability for the PWT system to detect other non-PWT applications with very short pulsed transmissions. The received signal strength indicator must be able to detect pulses of RF energy as short as 35 μs when they are at a specified signal level.

The FCC rules also limit the gain of PWT antennas to 3 dBi, whereas the DECT standard allows much higher gain values of 12 dBi or even 22 dBi with special permission.

2.4.3 Other Differences

Other differences between DECT and PWT occur in the MAC-layer protocol. Like the PHL, the MAC layer must support the extended channel plan. The portable also is expected to look for an additional MAC-layer message from the base station saying which carriers are in use. The MAC layer must enforce FCC etiquette rules that colocated devices may operate on no more than three carriers in total during any 10-ms frame, and it must use timeslots on an already active carrier if three carriers already are in use.

The following restrictions also are required:

- A 30 s limit is placed on independent beacon transmissions (those not associated with traffic) before the channel must be reassessed.
- A limit of 8 hours per call is imposed, after which a handover must be forced, to reevaluate the radio channel.
- During reassessment of a channel that previously was in use, a random deference period of between 10 and 150 ms (or a period greater than 150 ms) must be allowed before using the channel again.
- The systems must use a specified listen-before-talk (LBT) channel-selection process.

At protocol layers higher than the MAC layer, the voice applications profile, the CPAP, has several additional mandatory elements compared to the European GAP. At the application layer, the speech transmission levels are slightly different to accommodate the differences between European and United States telephone networks.

2.5 Interoperation Issues

DECT and PWT handsets, when built to an interoperation standard, are meant to work not only with the base stations that were designed with them but with other base stations designed independently, perhaps by another manufacturer. To achieve that, a number of implementation issues must be carefully dealt with. Some of those issues can be solved if the designers are aware of the need to guard against several potential future problems. They include, but are not limited to:

- Future changes in the standards;
- The need to support interoperation with equipment built under other possible interpretations of the standards;
- Passing the test suites.

Even if today you do not anticipate your handset operating with another manufacturer's base station or vice versa, you can be almost certain that some of your customers will try out such interoperation.

The big problem is that other manufacturers make independent decisions on implementation details. You also may have to continue working with new implementations coming onto the market. They may be your own new systems or those of other manufacturers, and they may be built to take advantage of new features in the standard. You may want to ensure that all your equipment in the field keeps operating when used with your new systems, so from your very first generation of equipment you need to be aware of design issues that will prolong its operational lifetime. If you do that, you will be able to exploit new system features without having to worry too much over whether your customers will be annoyed that their current handsets no longer work with your new systems.

2.5.1 The Robustness Principle

Many potential problems can be avoided if your system development teams adopt the *robustness principle* as basic practice. The robustness principle can be stated as the following: "Be conservative in what you transmit and be liberal in what you receive."

For many years, this principle has been behind the development of the independent networks (the intranets) that today are successfully interconnected to form the Internet. Just as a handset from one manufacturer may have to interoperate with some other manufacturer's base stations, the growth of the Internet has been achieved by many networks from different implementers being joined and successfully interoperating.

You can see the necessity for this principle in the situation where your handset must work on someone else's public base station. Your handset first of all must be liberal in what it receives—it must accept any legal transmissions from another manufacturer's base station. You do not know precisely how the base station's manufacturer will interpret the standard, they may implement differently some feature for which there may be more than one option. In that case, it is clear that you must be aware of all the options available to implement a feature you provide and you must be prepared to deal with any of them if you want the feature to always work.

The robustness principle introduces a requirement to understand fully the options available to the designers of DECT and PWT systems and not to dismiss them unless you are sure they are not permitted. In fact, it often is a good idea to take account of procedures that obviously are not permitted but that may be effective. Remember also that the protocol tester you are using will be a system with which you interoperate and almost certainly will exercise some of, but maybe not all, the options available when it is testing your equipment. Passing the protocol tests on their own may not be enough to guarantee that you have done a good job.

The robustness principle also makes it clear that you must differentiate between what you transmit and what you receive. Quite often, the standards make it clear what you must transmit without being so clear about what you should cope with on receive. In general, you should transmit precisely what the standards say, but you should be generous about what you receive, either ignoring things that you cannot interpret but that are not fatal errors, or acting properly if the information you have received is unexpected but sensible.

2.5.2 Some Specific Examples

What follows may seem obvious. However, the authors have come across specific instances of each of the "crimes" against the robustness principle, as illustrated by the following examples. Sometimes implementers failed to appreciate the possibility that implementations other than their own could operate in a different but permitted way. Sometimes they did not realize that in the future the standard would evolve and have more features added.

2.5.2.1 Reserved Bits

Reserved bits are the first example. Many of the fields defined in the higher layers of the standards contain reserved bits. The reason they are there is that the committees are reserving the bits for future use, so they may change in future implementations of both your own equipment and that of others with which you may have to operate. The proper rules for treating reserved bits are as follows:

- *Be conservative in what you transmit.* This rule is easy to obey. It simply means setting the bits in a reserved field that you are about to transmit to the values specified in the standard. Often, reserved bits are set to zero, but that is not always the case.
- *Be liberal in what you receive.* In the case of reserved bits, you must realize that in future editions of the DECT standard, reserved bits may be assigned to new functions, so their values must be ignored, not decoded. If your equipment relies on a reserved bit retaining its value, it may suddenly fail to interoperate in the future either with another manufacturer's equipment or, worse, with your own.

2.5.2.2 Reserved Values

Reserved values in fields that are otherwise specified usually are treated in the opposite way as reserved bits, despite the similarity in name. In this case, you have to decode the field on receive because all the currently specified values have meaning, and some may be meaningful in your implementation. The rules here are as follows:

- *Be conservative in what you transmit.* Do not transmit any of the reserved values specified in the standard. In the future, any "private agreement" you have set up between the handset and the base station you are designing may be misunderstood when a new system interprets the field according to a future standard. The new interpretation probably will not be what you want.

- *Be liberal in what you receive.* For reserved fields, this translates to the general principle that things you do not understand should be ignored and should not cause misoperation. Occasionally, the use of a previously reserved value in a field in one of the higher-layer protocol messages might be the cause for a system to be unable to continue because a critical piece of information is missing. That should happen rarely, if ever, and even then the systems should close down properly rather than crashing.

2.5.2.3 Message Lengths

Message lengths in the higher-layer protocols sometimes are specifically stated as variable. Future additions to both NWK-layer messages and the information elements they contain are to be expected in new editions of the standard and therefore in new equipment coming on the market.

- *Be conservative in what you transmit.* Do not use the principle that standard messages and their components can be extended for adding your own information. If you need to transmit private information, the protocol has mechanisms built in for proprietary extensions. Use those mechanisms instead.
- *Be liberal in what you receive.* Do not decode messages based on the specific message structures illustrated in the current standard. Use general principles to determine the actual length of the message you have just received so you know where the next one starts. Do that even if your implementation does not know how to decode all the information you have received, then just skip over and discard the new information you do not understand without allowing any misoperation.

2.5.2.4 Specified Alternatives

Specified alternatives exist within the standard for some procedures. Take the case where a feature has two possible alternatives in the standard that you may choose from. Even if you decide to use only alternative A in your systems, your implementations must deal with the possibility that they may operate with another system that chooses alternative B. The greatest dangers here are organizations that design and build all parts of a system themselves and the implementers are tempted to make choices for their own systems to the exclusion of other possibilities.

- *Be conservative in what you transmit.* Do not mix the methods; stick closely to the requirements of the standard.
- *Be liberal in what you receive.* Make sure you are able to work with all alternative methods of implementing a feature you want to provide, even if your own systems provide that feature in only one way.

References

[1] ETSI, *Digital Enhanced Cordless Telecommunications (DECT); Common Interface (CI)*, ETS 300 175 (9 parts), Sophia Antipolis, France, September 1996.

[2] TIA/EIA, *Personal Wireless Telecommunications Interoperability Standard (PWT)*, TIA/EIA 662-1997 (13 parts).

[3] Federal Communications Commission, *Unlicensed Personal Communications Service Devices,* CFR 47 part 15, subpart D.

[4] ETSI, *DECT General Terminal Attachment Requirements,* TBR 006, Sophia Antipolis, France, January 1997.

[5] ETSI, *Digital Enhanced Cordless Telecommunications (DECT); Generic Access Profile (GAP),* EN 300 444, Sophia Antipolis, France, August 1997.

[6] TIA/EIA, *Personal Wireless Telecommunications Interoperability Standard (PWT); Part 9: Customer Premises Access Profile (CPAP)*, TIA/EIA 662-9-1997.

[7] ETSI, *Radio Equipment and Systems (RES); Common Air Interface Specification To Be Used for the Interworking Between Cordless Telephone Apparatus in the Frequency Band 864,1 MHz to 868,1 MHz, Including Public Access Services,* I-ETS 300 131, Sophia Antipolis, France, November 1994.

Part II
Protocols and Implementation

3

Basic Protocols and Their Layering

3.1 Introduction

In the most positive sense, DECT and PWT are complex systems. Like ISDN, they are capable of providing many different services, features, and facilities. They offer the full microcellular operations of a digital cellular system (such as handover and support for location tracking) and provide the same level of security (including authentication and encryption). They can be used in several different applications, from a simple cordless telephone to a full microcellular wireless local loop system, and can interwork with digital cellular systems based on GSM.

DECT and PWT have been specified with specialized, highly structured methods and terminology. While being thorough (if tedious) to the initiated, the standards can be baffling and impenetrable to the newcomer. This section presents the structure of the protocol and introduces the methods used to describe and specify the system as well as some of the jargon. It is intended as an introduction to allow the reader to decipher the more detailed descriptions in subsequent sections.

The DECT and PWT standards specify only the air interface, those transmissions sent over the ether between the handset and the base station. That does not mean the standards can ignore what the equipment is connected to. The system must support a set of functions that are available in the network to which it is connected if it is provide signaling messages to control those features. For example, if the equipment is connected to a public switched telephone network (PSTN), then the signaling must support on-hook, off-hook, dialing digits using pulses or DTMF (dual-tone multi-frequency) tones, and sending DTMF tones during a call. If the base station is connected to an ISDN, then the signaling across the air must allow the handset to access and control a set of ISDN features (although perhaps not all). If the base station is connected to a GSM network, then the signaling must provide access to a set of GSM features, such as location updating, authentication, and short messages.

DECT and PWT have been specified with methods that are now common practice for modern standards. To manage the complexity, the standard is broken down into more manageable modules. The modules are defined in terms of logical functions rather than in terms of physical equipment, thus allowing manufacturers the freedom to choose their preferred implementation.

The first step is to determine the main elements of the network, what each element does, and thus the location and functions of the interfaces. The list of features that the signaling system must support is defined by features needed with the cordless system itself (e.g., handover) and the functions in the associated networks (e.g., dialing a telephone number). Rather than defining every conceivable physical network to which DECT and PWT could be connected, the signaling system is defined by a setting up of the cordless universe in terms of major functional elements, the portable part, the fixed part, the interworking unit (IWU), the local network, the home location register, and so on. It is a generalized representation of many different real physical networks. While the interfaces between those elements are not specified, the cordless universe does provide the standards designers with a record of all the functions that the signaling must support.

After determining the main network elements, the interface between the handset and the base station is defined by a hierarchy or "stack" of signaling protocols. Signaling at the top of the stack corresponds to high-level commands such as "set up a call to telephone number 897-555-0123." The lower layers of the hierarchy are responsible for breaking down the message into smaller pieces that can be transmitted reliably across the radio interface. Many types of signaling information have to be transmitted (e.g., call set-up messages, information on which radio channels are in use at a base station, information on which services are available at a base station, handover requests, requests to repeat incorrectly received messages), and all those different signaling messages must be organized so that each signaling message gets to the right place and causes the correct action to be taken.

Formal methods are used to provide an unambiguous and consistent method of description, which minimizes the risk of incompatibilities in the protocols. The highly structured and rigorous description is more than what would be needed to specify a simple cordless telephone, but it is an excellent basis for advanced services and features. (Note that the complexity of the standards is due in part to a lot of material that is relevant only to advanced applications.)

Signaling must be defined with both static and dynamic descriptions. The static descriptions provide definitions of messages, and the dynamic description describes the sequence in which messages can be exchanged. The standards documentation often follows that pattern, with the first part defining concepts and the latter parts giving the signaling procedures.

3.2 Functional Elements

DECT and PWT are intended to be used in may different applications, from simple cordless phones to full public microcellular networks. The standards try to facilitate as many options as possible without burdening simple products with features that are necessary only for exotic applications. Unlike, say, GSM, where a base transceiver station (BTS) is always connected to a base station controller and the BTS in turn is connected to a mobile switching center (MSC) and so on, DECT and PWT can be configured in many different ways.

3.2.1 The Overall Architecture

Some of the many potential DECT/PWT configurations are illustrated in Figure 3.1. The simplest case is a residential cordless telephone. Alternatively, DECT or PWT can be used to provide a multicell wireless PABX, which might be connected to PSTN, ISDN, GSM, or X.25 public networks. The handset might provide a "cordless socket" or RLL, into which a regular telephone or fax machine can be plugged. DECT or PWT also can be used to create a PC-card wireless modem, providing a personal computer (PC) with a high-speed wireless connection. It also can be used to provide a full standalone public microcellular network, offering, incoming calls, authentication of outgoing calls, and handover.

For the purpose of designing and specifying the signaling protocols, rather than considering all the possible physical configurations, the entire network of which the cordless system is a part is defined in the DECT/PWT reference model illustrated in Figure 3.2.

The overview in Figure 3.2 shows the relationship of the cordless system with the networks to which it is attached and illustrates that the networks are not part of the standard. That can be seen more clearly from Figure 3.3, which shows that the cordless system is connected to the network via an interworking unit. There are different IWUs for each different type of network (PSTN, ISDN, GSM, data networks, etc.). Those IWUs are not part of the core

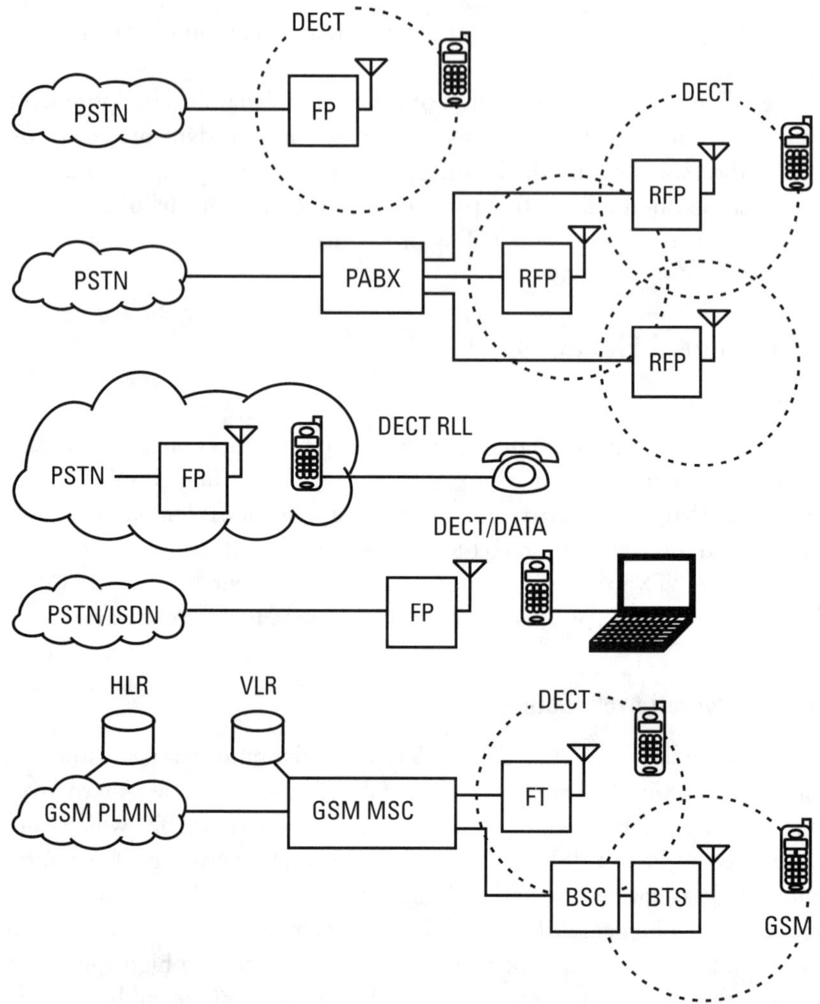

Figure 3.1 Some DECT/PWT configurations.

standards. It is usually straightforward to map any particular physical architecture onto the functional architecture, that is, to map which physical equipment performs which function.

As illustrated in Figure 3.3, cordless equipment consists of only two parts, the portable part and the fixed part, which are generalizations of a handset and a base station. Technically, the standard itself covers only the fixed and portable terminations that are connected by the radio interface.

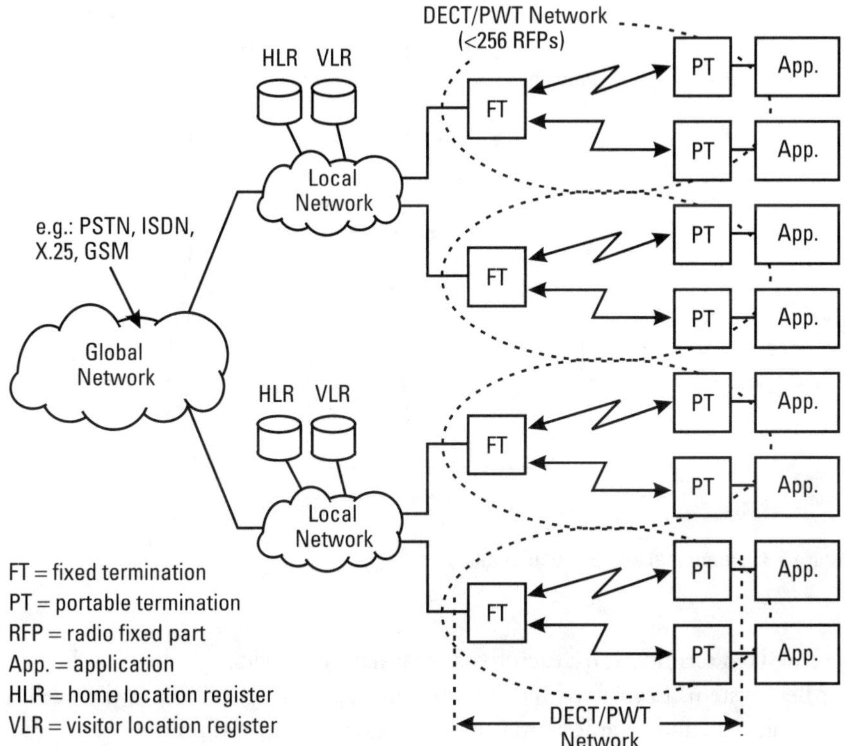

Figure 3.2 The DECT/PWT reference model.

3.2.2 The Portable Part

The portable part comprises the portable termination and the portable application or an end system. The portable termination provides all the cordless access functions, and the portable application provides all other functions on the portable side. That could be as simple as the handset's keypad and display or a more complex PC-card application designed to send faxes over a cordless link. If DECT or PWT is used to provide a cordless-socket type of product, it becomes an intermediate and transparent medium through which the end system obtains access to the network.

3.2.3 The Fixed Part

The fixed part could be anything from a self-contained residential base station to a wireless PABX or an entire microcellular network. A fixed part can contain

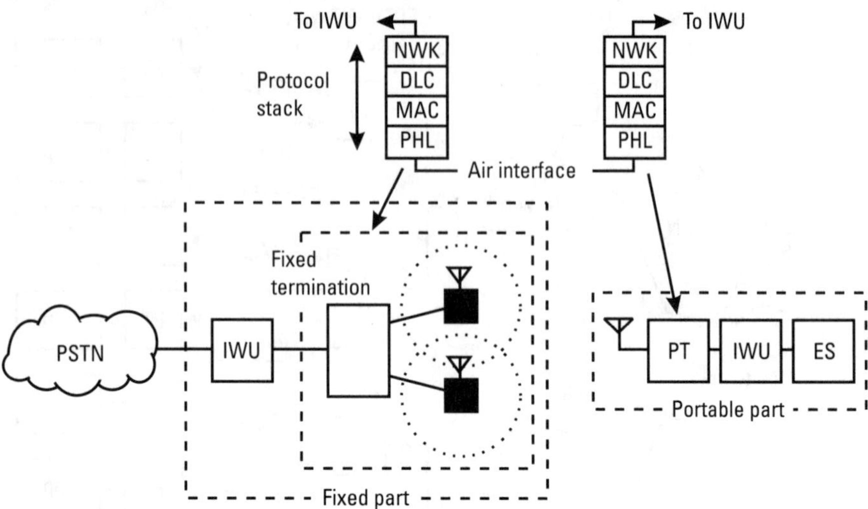

Figure 3.3 Functional and protocol entities.

several fixed terminations; each fixed termination provides all the functions of a cordless system. Each fixed termination may contain several RFPs (i.e., cells or base stations), and each RFP may contain several radio end points, that is, radio transceivers.

3.2.4 Global and Local Networks

The global network in Figure 3.2 is a telephone network (e.g., PSTN or ISDN), a data network (e.g., X.25), or a mobile network (e.g., GSM) to which the portable part can be connected.

The local network could be a PABX, a LAN, a multiplexer, or anything up to an entire private telecommunication network. It connects the cordless unit to the global network. It also connects calls between handsets, since DECT and PWT themselves do not provide any switching functions.

The two data bases, the HLR and the VLR shown in Figures 3.1 and 3.2, are used to direct incoming calls to mobile users. If a user is not always in the same place, that is, not in the same cell, it is necessary to keep a dynamic record of the user's current location. A single HLR is assigned permanently to each user and keeps a record of which VLR area (a group of cells) the handset currently is in. The HLR also can be used to store other personalized data such as PINs for validating call requests, information on whether the user has forwarded calls, or instructions that the user is barred from making certain classes of call.

3.2.5 The Interworking Unit

The cordless access system is connected to each different type of network via a different IWU. That means a cordless system can (in principle) be connected to anything by changing the IWU. Note, however, that the IWU is not magic; it can convert signals from one format to another, but it cannot interface DECT or PWT to a system that provides entirely different features and facilities.

The IWU converts the signals and messages used on the air interface to a format suitable for the network. For example, if a cordless base station is connected to a PSTN, the NWK layer will convert the call control message, {CC-SETUP} (asking for a call to be established), into an equivalent standard primitive to the IWU, which will then take application-specific action to take the telephone line off-hook (the term *primitive* is explained in Section 3.3.1). If the message contains a number to be dialed, the IWU will tell the telephone line interface to dial it. It also will convert the analog speech signal used on the PSTN to ADPCM used on the air interface, and vice versa. Similarly, if a data call rather than a voice call is selected, the IWU will utilize one of its modems to convert the digital signals received from the protocol stack to analog modem tones. When connected to an ISDN, the IWU will translate ISDN messages into equivalent DECT/PWT messages, or it simply may pass the ISDN signaling transparently across the cordless interface and allow a device on the other side of the cordless system to decode and act upon those messages. It also may adapt the data rate (e.g., insert null data) to match one of the standard data rates used on the air interface to the ISDN-line interface. When connected to a GSM network, the IWU must insert some of the GSM identities and messages into the cordless protocol's signaling and transport that signaling over the air interface to the GSM subscriber identity module in the handset.

3.3 Protocols, Layers, and Planes

The signaling between the fixed part and the portable part is complex. Therefore, it is defined in terms of more manageable modules, or "layers." That method allows the signaling to be conveniently divided into lower layers, which deal with low-level tasks (e.g., transmitting and receiving bits); higher levels, which deal with ensuring that signaling blocks are transmitted without error; and the highest layer, which deals with sending signaling messages such as "go off-hook," "dial 897-555-0213," "go on-hook," or "I am now present in this area; please deliver my calls here." The layers are specified with a formalized terminology that provides a consistent description of how the signaling should behave.

3.3.1 Layers and Planes

The DECT/PWT interface protocol is based closely on the ISO open system interconnection (OSI) layered model and in fact is specified using a four-layer model that comprises the following layers:

- The NWK layer;
- The DLC layer;
- The MAC layer;
- The PHL.

Those four layers actually cover the first three formal ISO layers. The PHL corresponds to ISO layer 1, the MAC is layer 2a, the DLC layer is layer 2b, and the NWK layer corresponds to layer 3.

This hierarchy of signaling protocols is commonly referred to as the protocol stack. Each layer of the stack provides the layer above with services or with a more abstract view of the physical medium over which protocol-specific messages, network-specific messages, and user traffic are transmitted.

Layers do not necessarily have any physical counterpart. They are used as a method to describe and define the externally observable behavior on the interface. In particular, each layer prepares and formats signaling messages for its "peer" layer in the handset or base station. For example, the handset's MAC layer protocol prepares and transmits messages that are understood by the base station's MAC layer protocol. Signaling messages are exchanged between peer entities within the same layer, and each layer provides a different protocol.

The interface between layers is defined in terms of what are called *primitives.* Primitives do not have any physical significance; their only use is to provide a formal method of describing how the layers interact with each other and thus how the signaling will behave on the air interface.

The primitives are passed between the layers at points called service access points (SAPs). It is useful to view a SAP as an address for a signaling message. Data are passed in the primitive in units called service data units (SDUs). In DECT and PWT, the control channel SDU length is constrained by the fact that in simple systems the control data multiplexing method permits only 40 bits of signaling to be transmitted in each frame. To transmit an entire message in those small units of data, the message has to be broken down to smaller pieces by segmentation and fragmentation, then recombined and reassembled at the other side of the link.

Layering concepts originally were developed for—and are still most suitable for—describing data or signaling applications. Transmission of speech signals also is described with the same layering method, but since the transmission of a speech signal is a simple regular speech coding and multiplexing process, there is less benefit in a description technique intended to simplify the description. Thus, there are different layered descriptions of the signaling and the speech signals, referred to respectively as the control plane, or C-plane, and the user plane, or U-plane.

Layered descriptions need one further element to describe radio systems adequately. In wireline data and voice systems, the signaling and user transmission mostly can be described by interactions between a layer and those immediately above and below. In most radio systems, however, several things do not fall into neat layers. For example, to choose a free radio channel, the higher layers must measure the strength of any radio signal on that channel, thus bypassing the normal layering mechanisms. Those interactions are handled via the lower-layer management entity (LLME). The complete protocol architecture is illustrated in Figure 3.4. In the example above, the LLME instructs the PHL to measure a particular channel, and the PHL returns the value. The LLME then may produce an ordered list of which beacon signals are strongest or which channels have the lowest interfering signal levels. This description does not

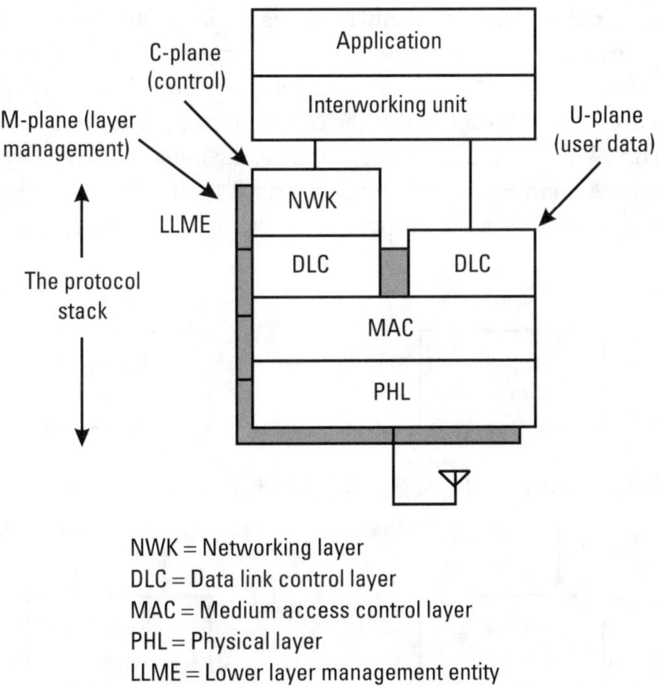

Figure 3.4 The layered protocol architecture.

require that the product contain distinct modules performing the distinct tasks, rather, it simply means that the channel has to be measured. Note that the IWU interfaces to the top level of protocol stack and does not normally interact directly with the LLME.

3.3.2 Layer-to-Layer Communications

Layered air interface protocols, such as DECT and PWT, usually try to hide what goes on at their lower protocol layers from processes at higher layers. That is typical in almost all layered protocols and is about the only way to make the complexity of operation tractable. It usually is achieved by employing purpose-designed messages between the layers of a protocol, called *primitives*, which pass both up and down the protocol stack. The use of primitives is a formal method for making it easier to describe what happens; they are used later in this book and throughout the standards to describe how certain procedures work. You do not, however, have to implement them precisely as given, or at all.

Use of primitives usually follows a specific pattern based on four types, called *request*, *indicate*, *response*, and *confirm* (shortened to req, ind, res, and cfm). A typical operation is illustrated in Figure 3.5.

In Figure 3.5, the target's initiator's higher layer initiates an action for its lower layer (e.g., establishing a connection to a target) with a LAYER_ACTION-req primitive (point 1). A real example in DECT and PWT is the DL_ESTABLISH-req primitive that the NWK layer sends to the DLC layer to get it to open up a link. On receipt of the primitive, the initiator's lower layer conducts some communication through the protocol stack (point 2), entirely hidden from the higher layer (which is the purpose of layering). When the target receives an

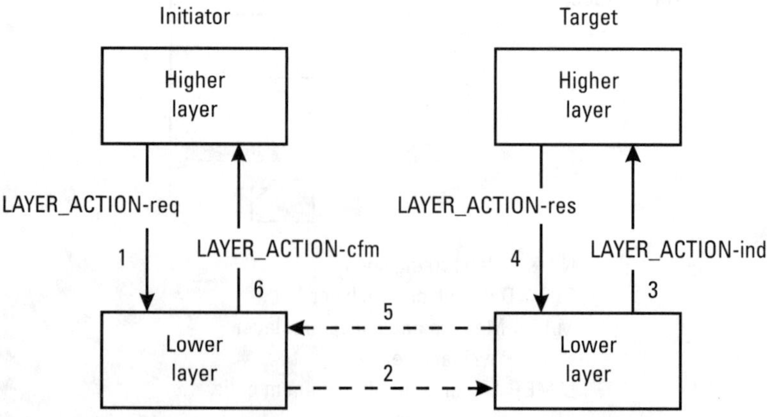

Figure 3.5 Operation of primitives.

indication of what has happened at the initiator, it indicates to its upper layer with a LAYER_ACTION-ind primitive that something unexpected has happened (point 3), something not requested by the target. In Figure 3.5, the lower layer requires its upper layer to respond to that primitive with a LAYER_ACTION-res (point 4). Now some more hidden communication takes place within the protocol stack (point 5), before the initiator's lower layer finally confirms to its higher layer with a LAYER_ACTION-cfm that the requested action has been carried out (point 6).

The -ind and -cfm primitives are sent only from lower layer to higher layer, and the -req and -res primitives go only from higher layer to lower layer. The -ind primitive indicates that something unexpected or unrequested has happened, and the -cfm primitive confirms a request. The -req primitive normally is a request for a lower layer to do something, and the -res primitive is the response that it has been done. Note that not all the primitives are used in all cases. For example, the LAYER_ACTION-res (point 4 in Figure 3.5) may not be required by procedures at the target, and its lower layer may be mandated to notify its higher layer and autonomously initiate any confirming action. Furthermore, an action may be initiated by the lower layer for the higher layer, for example, to request data from the higher layer (the MAC layer does that to get user traffic from the DLC layer). In that case, perhaps slightly counterintuitively, the action is initiated by a LAYER_ACTION-ind, and the response is a LAYER_ACTION-req.

Primitives usually contain some data. The MAC layer sends a PL_TX-req primitive to the DLC layer, which contains the data packet to send and, among other things, an indication of the channel on which to send it.

The use of primitives for some actions, for example, data transfer, probably will not introduce the concept of states into the protocol. However, the use of primitives for opening and closing connections will introduce states into the lower layer. It may be that data transfer cannot take place with the lower layer in the closed state but can in the open state. An example is given in Figure 3.6, where two actions, the establishment of a connection and the release of a connection, act together to introduce the concept of an open-connection state and a closed-connection state in the lower layer.

Because the example in Figure 3.6 requires cooperation between initiator and target for both opening and closing the connection, two additional states are introduced into the lower layer as seen from above—the open pending and close pending states.

The transition from the closed state to the open state goes first to the open pending state as a result of sending the lower layer an ESTABLISH-req and then to the open state on receipt of the ESTABLISH-cfm. In the case of a failure of the lower layer to complete the requested action, it responds with a

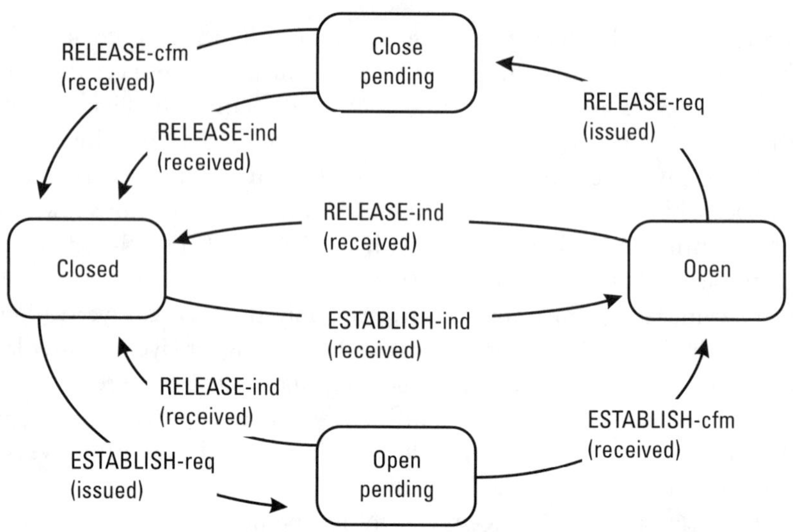

Figure 3.6 The states in the lower layer, as seen from above.

RELEASE-ind instead, taking the state back to closed. Such a failure may be as a result of the lower layer having insufficient resources or may be due to the failure of the target to respond within a timeout period. In the case of timing-related issues, it usually is advantageous for the lower layer to take responsibility. If for any reason a particular primitive cannot be acted on in a specified period, the lower layer will always respond negatively to the higher layer, a principle that simplifies the higher layer's operation.

The example here is simple compared to a real cordless system. For instance, it does not deal with all possible error conditions. However, the principles are used extensively by DECT and PWT (and also later in this book) to describe complex layered protocols in understandable ways.

3.3.3 Functions of the Network Layer

Figure 3.7 gives an overview of the use of the layered structure to send and receive signaling messages. It shows the type of data exchanged between peer protocol entities and the interactions between layers.

The NWK layer is the "executive" layer. It interfaces to the outside world and relies on the lower layers to set up a path so it can communicate with the NWK layer in the other cordless part. In the portable part, it may, for example, receive an indication from the IWU that a handset has gone off-hook and dialed a number. The NWK layer will instruct the lower layers to set up a

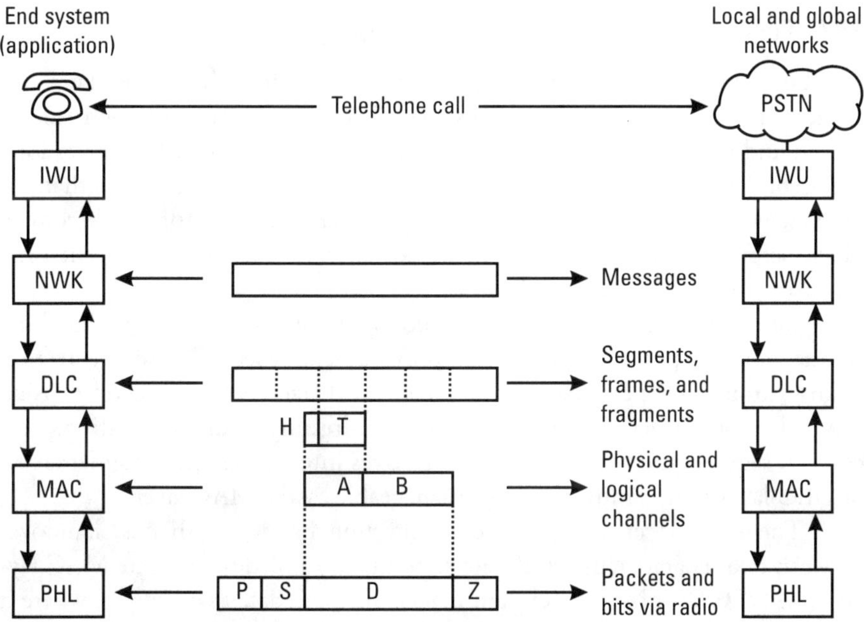

Figure 3.7 Layered signaling.

channel and (implicitly) transport the call setup message to the NWK layer protocol entity in the fixed part. That will decode the instruction and instruct the PSTN IWU to take the telephone line off-hook and dial the number specified.

The NWK layer exchanges *messages* with it peer NWK layer. Messages are composed of information elements that are commonly used pieces of information, for example, the identities of individual handsets and base stations or dialed numbers. It also provides separate functions for the following:

- Call control (CC) (setting up and clearing calls);
- Supplementary services (SS) (diverts, transfers, etc.);
- Connection-oriented message service (COMS) (sending messages only when an explicit channel has been set up to the receiver);
- Connectionless message service (CLMS) (sending messages without an explicit channel set-up);
- Mobility management (MM) (authentication, location registration, and handover).

3.3.4 Functions of the Data Link Control Layer

The DLC layer provides reliable signaling (or data) links. It can be likened to an accounts department that checks that all the messages go though uncorrupted and in the correct sequence. It is purely a middle man. It receives long blocks of data from the NWK layer and breaks them down into blocks that can be managed by the lower layers. The DLC layer adds an error check/control block before passing the data on to the lower layers (using what is called the link access protocol type C, or LAP-C).

The DLC layer can provide both point-to-point and broadcast signaling. In the point-to-point mode, it can provide acknowledged error correction (using automatic repeat requests) or unacknowledged error protection (using forward error correction). It also provides addressing, frame delimiting, flow control, segmentation of NWK layer messages into frames, and fragmentation of DLC-layer frames into fragments manageable by the MAC layer.

The DLC layer also provides connection handover, that is, handover between two separate clusters of base stations, controlled by separate MAC layers. The DLC layer is more relevant to signaling and data than speech, where it has no real role. However, to keep the description terminology consistent, it is considered to provide a "transparent unprotected service" for voice connections. For data services, however, the DLC layer provides a family of alternative services, including:

- Frame relay service;
- Frame switching service;
- Rate-adaptation service.

3.3.5 Functions of the Medium Access Control Layer

The MAC layer creates traffic and signaling channels out of the raw data packets provided by the PHL. It also divides the available signaling capacity between different uses, such as paging and call setup messages. The MAC layer creates traffic channels by instructing the PHL when to transmit and when to receive. It is also responsible for broadcasting the beacon signal that gives a handset all the information it needs to set up a call to the correct base station in the correct channel.

The MAC layer controls the bandwidth of the radio channel and can create both simple narrowband channels and broader-band "multibearer" channels. It is responsible for establishing and maintaining connections on those channels. A single MAC layer can control a group of up to 255 base stations, called a cluster, and it provides a bearer handover between base stations.

The physical channel has only a limited signaling capacity, and the MAC shares that among different uses. Different types of messages are grouped into logical channels, which the MAC multiplexes, with a defined set of priorities, into the available signaling channels. The MAC then specifies how the signaling is mapped onto the physical channels.

3.3.6 Functions of the Physical Layer

The PHL does little more than gain and maintain synchronization with its peer and then transmit or receive packets of bits. The PHL specifies radio parameters: frequency, timing, power, bit and slot synchronization, modulation, and the performance requirements for transmitter and receiver. The MAC layer and the LLME tell the PHL what to do and when to do it. Its only autonomous actions are to check for and report synchronization slippage.

3.3.7 Practical Realization

Figure 3.8 shows a fairly typical set of practical cordless subsystems and illustrates the position of the various layers.

In a reasonably typical cordless PABX implementation, a line card may be a complete fixed termination that connects the protocol stack to the PABX via its internal IWU. Normally, the PHL resides in remote RFPs that are at the center of each radio cell. The mobility database is an application that uses the protocol stack and is therefore a part of the PABX application.

In the handset, the burst-mode controller chip implements the lower parts of the MAC layer, leaving the upper parts of the MAC layer, all higher protocol layers, and the handset application in the microcontroller and specialist modules. The PHL often is in a separate module of its own.

3.4 Base Standards and Application Profiles

We have already seen in Chapters 1 and 2 that DECT and PWT systems are specified in base standards [1,2], which are largely application independent, plus application-specific profiles for systems providing certain services, for example, voice communications [3]. However, it is a little more complex than that. There also are several technical bases for regulation (TBRs), which are ETSI's technical parts of the European Union's common technical regulations (CTRs) that in Europe make some parts of the DECT standards mandatory, such as the radio interface [4], voice telephony [5], and voice service profiles

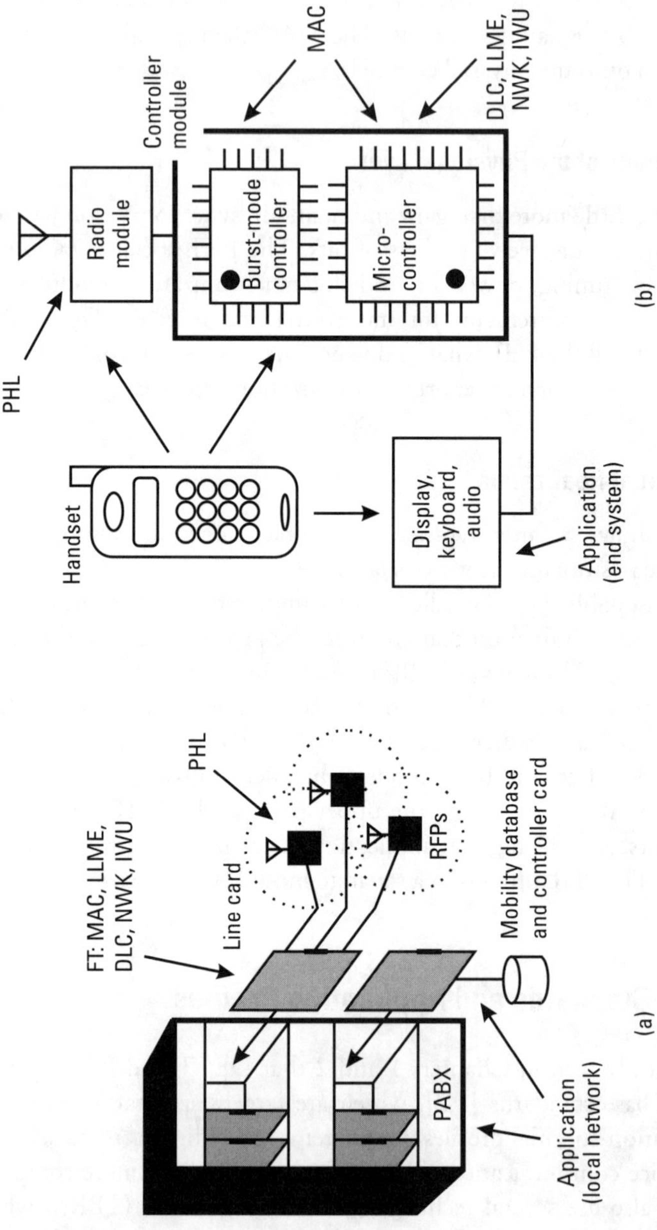

Figure 3.8 The protocol layers in (a) a PABX and (b) a handset.

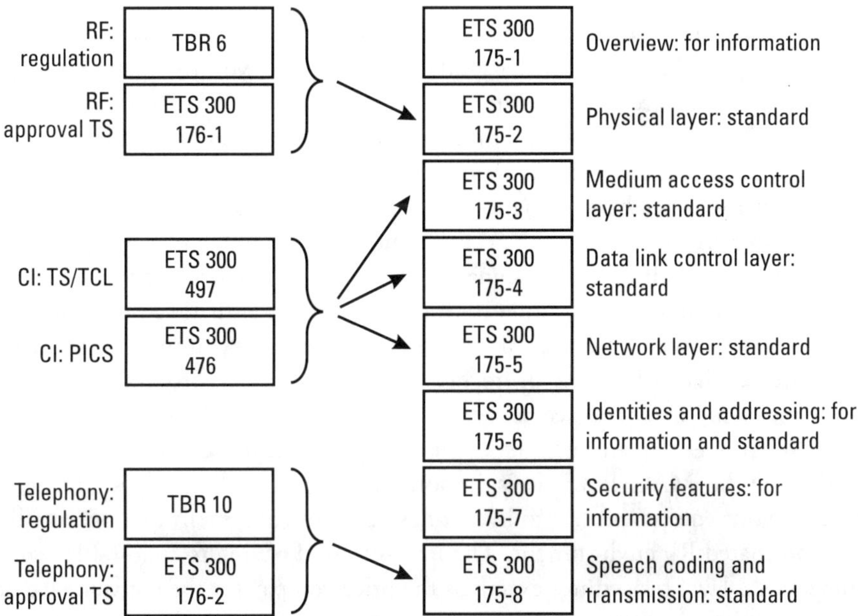

Figure 3.9 The DECT base standards and their test specifications.

[6,7]. On top of all that, there are test case libraries and profile implementation conformance statements (PICS).

This section examines details of the standards and the relationship between the standards and the regulations.

3.4.1 The Base Standards

Just like the protocol itself, the DECT and PWT base standards are layered according (at least approximately) to the ISO layering model. Figure 3.9 illustrates the standards for the protocol stack: the PHL (the radio), the MAC layer (how to access the PHL), the DLC layer (how to guarantee data integrity), and the NWK layer (how to set up calls and get other cordless communication services). A number of related standards support the protocol stack, including testing, security, and system identities.

The base standards describe procedure and data flows for a whole range of applications; without any further guidance, they invite you to "pick and mix" to achieve the features you want for a particular product. For example, a handset that is intended for use on a simple cordless phone does not need to support the signaling needed to communicate with, say, a GSM network or a data network. On the other hand, a handset that is intended to access GSM services and features via the air interface needs to know exactly which messages the base station will transmit and what signaling responses will be expected of it. Actually, each item on the base standard's menu also comes with a recipe for how to do it. So if you pick a feature, the recipe is specifies how you must implement the feature. That is how the protocol's provision optional, process mandatory principle is implemented (see Section 2.1.3).

What does that mean? First of all, at the lower layers of the protocol, the PHL and the MAC layer, it is important to specify the recipes precisely, because your equipment has to live alongside other equipment in a potentially uncoordinated RF environment. The imposition of having to follow PHL and some MAC-layer procedures exactly is the price you pay for the convenience of having minimal interference to your equipment from anyone else's operating in the same geographical and spectral area. Specifying the lower-layer procedures in great detail provides you with coexistence. Actually, pure coexistence between colocated systems does not require complete adherence to the full MAC layer. At the MAC layer, it is possible to follow only the first sublayer of MAC structure and procedures, which will allow another system to work in cooperation with your system to allocate and use RF resources and to recognize your system as one that does not follow the specified protocol at any higher layer. You will find that the MAC layer specifies an escape mechanism to a completely proprietary upper MAC protocol that still will allow you cooperative access to the radio spectrum used by other DECT and PWT systems. If you take the base standards only, the higher layers of the MAC layer, the DLC layer, and the NWK layer all contain no more than suggestions of how to operate your system. If only the base standards existed, only coexistence of differing DECT and PWT systems could be achieved.

3.4.2 Voice Application Profiles

In certain applications, the concept of interoperation is either important or desirable. Take, for example, a *public access* telephone service, sometimes called *telepoint.* A telepoint service using a network of base stations is rather like having a network of public telephones in which everyone has a handset and can plug it in. The managers of this sort of service may want to base their business plans on the existence of many residential cordless telephones, whose owners are potential subscribers to a public access system. More than likely, those telephones will be produced by a number of different manufacturers. In those conditions, a standard for the air interface that specifies the operation of the cordless telephone right up to the top of the NWK layer is needed to ensure that anyone can connect a cordless handset to the telepoint network. Of course, it does not need to specify every feature in order to work. Only those features supporting the level of service desired need to be standardized. That gives rise to the concept of the *application profile* for each specific service (Figure 3.10) and explains why so much trouble has been taken to specify a rich set of features in the base standards and not force everyone to use them.

The application-specific profiles, such as the GAP [3], specify which parts of the base standards have to be implemented in which part of the system to guarantee specified service with independently manufactured fixed and portable equipment. The key parts of a profile are as follows:

- A scope defining which applications are covered;
- A list of the features and procedures from the base standard that are mandatory, optional, or not applicable, specified for each part of the applicable system(s);

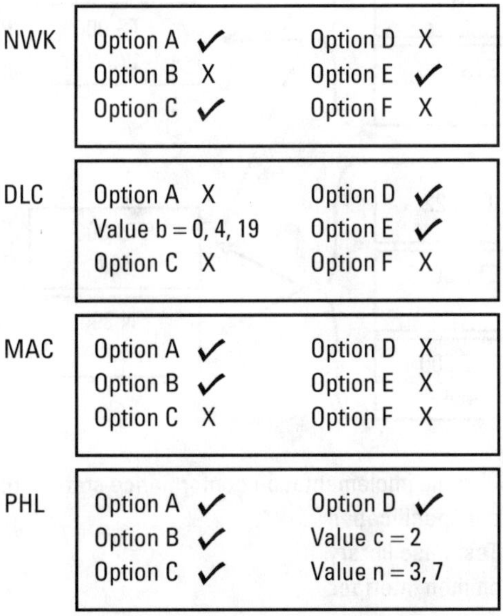

Layer		
NWK	Option A ✓	Option D X
	Option B X	Option E ✓
	Option C ✓	Option F X
DLC	Option A X	Option D ✓
	Value b = 0, 4, 19	Option E ✓
	Option C X	Option F X
MAC	Option A ✓	Option D X
	Option B ✓	Option E X
	Option C X	Option F X
PHL	Option A ✓	Option D ✓
	Option B ✓	Value c = 2
	Option C ✓	Value n = 3, 7

Figure 3.10 An application profile is a subset of features and procedures.

- A blank form (a profile implementation conformance statement or PICS) that allows a manufacturer to record which features of the base standard are implemented in the equipment.

For the telepoint application, ETSI created the Public Access Profile (PAP). That standard was the first application profile, but now it is obsolete and replaced in Europe by the GAP [3] and in the United States by the CPAP (part 9 of [2]). The new speech profiles cover not only public access telephony applications such as telepoint but all applications that are capable of carrying voice. That includes residential cordless telephones and business PABXs.

The DECT GAP is mandatory in Europe for all voice-capable applications; hence, legally binding regulations have been produced based on ETSI standard TBR 22 [7]. The PWT CPAP does not have the same regulatory status. However, its purposes are the same: to allow equipment from different manufacturers to interoperate and to provide basic voice service. The relationship among the two voice service application profiles, the TBRs, and their test specifications is illustrated in Figure 3.11.

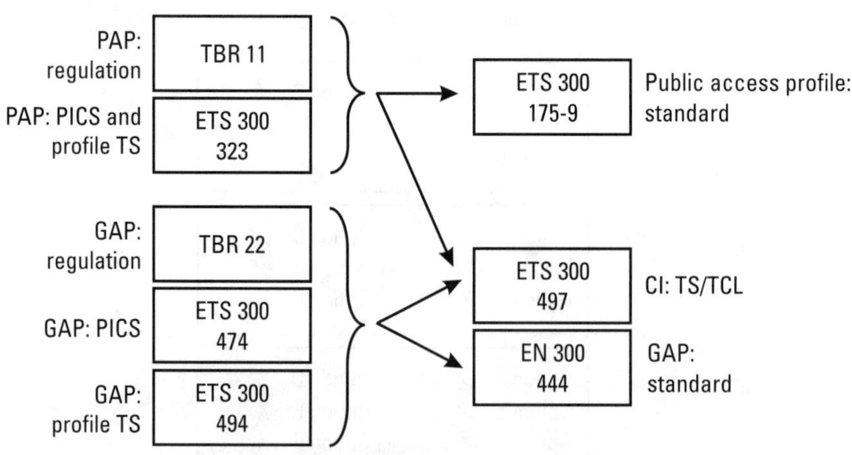

Figure 3.11 DECT voice service application profiles.

3.4.3 Test Specifications and Implementation Conformance Statements

The base standards are backed by a test case library, ETS 300 497, which specifies protocol tests for all the features in the MAC, DLC, and NWK layers of the protocol.

The test case library is accompanied by an implementation conformance statement, ETS 300 476. That is simply a table that a manufacturer fills in to say which of the base standard's features are implemented in a piece of equipment. It usually is used as either a statement to an independent test house about which tests to apply to the equipment or a record for the manufacturer's own use in putting together a test plan. When no regulatory requirements are imposed, the implementation conformance statement is optional. That is certainly true of the base standard implementation conformance statement. However, when it comes to application-specific profiles that are the subject of regulation, the manufacturer usually has to fill in the specific PICS either to gain a positive test result from an external test house or to meet legal obligations to maintain internal test records.

3.4.4 Proprietary Protocol Options

As well as providing highly prescribed procedures about how handsets and base stations should work together, the standards also permit advanced or experimental features through the use of proprietary signaling messages. In applications where there is no application profile forbidding it, DECT and PWT permit proprietary signaling to be added at each of the major signaling layers, the MAC, DLC, and NWK layers. The biggest escape point is just at the lower part of the MAC-layer protocol. Operation according to the standard up to that point is essential for systems to coexist without mutual interference, but you can build a compete new signaling protocol on top of the coexistence standard.

It should be noted, however, that if you are using the standard signaling and want to add a feature in a proprietary way, you need to check if there already is a standardized way of performing that particular feature. If so, the standardized method must be used, not an equivalent proprietary method. It is not possible to mix and match proprietary signaling, and once signaling has "escaped" to proprietary it cannot successfully revert to standardized signaling.

3.4.5 Other Application Profiles

The base standards specify as far as they can how to connect a cordless system to an analog PSTN and a digital ISDN if you do not need any ISDN-specific features, while leaving open the means to achieve specific national requirements. However, if you want to interwork properly with ISDN, ISDN interconnection can be provided in two ways: the *end-system* and the *intermediate-system* approaches. The end-system ISDN interoperation provides access to ISDN services for the portable itself. The intermediate system provides for a socket to be present on a portable into which a fixed ISDN terminal may be plugged, giving it access to the ISDN over a cordless link.

The interworking of DECT or PWT with GSM-class digital cellular systems occurs on two levels. In the first level, a cordless fixed part is attached to the A-interface of a GSM system, and the handset is made to appear to the cellular system as though it is just another cellular handset. At the second level, a cordless PABX may be connected to a cellular system via a 2-Mbps ISDN link. The technical details of the interworking of cordless and cellular systems are a specialized topic covered specifically in Section 8.5. There usually are important commercial details that need to be sorted out to allow a cordless system to connect to a cellular system, even if the technical matters are resolved.

The increasing importance of data communication has led to a complete framework for cordless data applications covering frame relay, data streams, point-to-point protocol (PPP), messaging, and mobile multimedia applications. More details on those interworking applications are presented in Section 8.6.

3.4.6 Other Important Standards

All the DECT and PWT standards are listed in the bibliography at the end of this book. Some other important standards are listed in Table 3.1.

The first two standards in Table 3.1 set out requirements for generic radio systems to meet the European Union's directive on EMC [8] and the

Table 3.1
Other Important Standards

Standard	Contents
ETS 300 339	A generic standard for radio system EMC
ETS 300 329	The specific EMC standard for DECT radios
ETS 300 331	A DECT authentication module (DAM), a smart card for storing subscriber information
ETS 300 700	DECT wireless relay stations (protocol repeaters to extend radio range)

specific requirements for DECT systems [9]. Meeting those requirements is mandatory in Europe to be able to put the mark "CE" on DECT equipment.

Later in this book, we will see that data have to be stored in a handset whenever it is subscribed to a telecommunications service; some of the data allow the handset to recognize the service, and some allow the service to securely recognize the handset. If the data are associated with memory fixed within the equipment, it is not possible for the user to use any other handset to gain access to subscribed services. However, a DECT authentication module (DAM) is specified [10], which is a smart card that can securely carry the user's subscription data and that can be swapped from handset to handset to allow the subscription to move freely from equipment to equipment.

The fourth standard in Table 3.1 is for how to build a wireless repeater [11]. Such a device relays cordless transmissions between a handset and a base station that otherwise are too far apart to communicate. In some territories, where the regulatory authorities permit, chains of repeaters may be used. However, the amount of spectrum used for a single communication increases with the number of repeaters, and often no more than one repeater hop is permitted.

References

1. ETSI, *Digital Enhanced Cordless Telecommunications (DECT); Common Interface (CI),* ETS 300 175 (9 parts), Sophia Antipolis, France, September 1996.

2. TIA/EIA, *Personal Wireless Telecommunications Interoperability Standard (PWT),* TIA/EIA 662-1997 (13 parts).

3. ETSI, *Digital Enhanced Cordless Telecommunications (DECT); Generic Access Profile (GAP),* EN 300 444, Sophia Antipolis, France, August 1997.

4. ETSI, *DECT General Terminal Attachment Requirements,* TBR 006, Sophia Antipolis, France, January 1997.

5. ETSI, *DECT General Terminal Attachment Requirements: Telephony Applications,* TBR 010, Sophia Antipolis, France, January 1997.

6. ETSI, *Attachment Requirements for Terminal Equipment for DECT Public Access Profile (PAP) Applications,* TBR 011, Sophia Antipolis, France, September 1994 (with amendment of March 1995).

7. ETSI, *Attachment Requirements for Terminal Equipment for DECT Generic Access Profile (GAP) Applications,* TBR 022, Sophia Antipolis, France, January 1997 (with amendment of November 1997).

8. ETSI, *General Electro-Magnetic Compatibility (EMC) for Radio Equipment,* ETS 300 339, Sophia Antipolis, France, June 1997.

9. ETSI, *DECT Electro-Magnetic Compatibility (EMC) for DECT Equipment,* ETS 300 329, Sophia Antipolis, France, June 1997.

10. ETSI, *DECT Authentication Module (DAM),* ETS 300 331, Sophia Antipolis, France, November 1995.

11. ETSI, *DECT Wireless Relay Station (WRS),* ETS 300 700, Sophia Antipolis, France, March 1997.

4

The Physical Layer

4.1 Introduction to the Physical Layer

The radio interfaces [1–3] are the foundations on which DECT and PWT are based. Despite the fact that the PHL description is one of the shortest and ostensibly the simplest, it is the PHL that gives DECT and PWT many of their key characteristics.

Stated simply, the PHL provides transport of data packets from one end of the radio link to the other. The higher layers of the DECT protocol make "bit pipes" by regularly getting the PHL layer to transmit and receive data packets (Figure 4.1), and use those bit pipes to carry telephone conversations and other services.

To share the available radio spectrum among many users, the PHL divides the spectrum into individual physical channels. It does that by designating radio carriers within the operating band and the means to modulate digital data onto them. There is a common definition for a multicarrier TDMA scheme based on a 10-ms frame having 24 timeslots, with a common scheme for transmitting and receiving data packets of various lengths at 1,152 kbps within one or more of those timeslots. The PHL also specifies radio parameters such as frequency accuracy, timing accuracy, RF power levels, bit and slot synchronization, transmitter requirements, and receiver performance. Ultimately, the PHL outlines everything needed to allow independent systems to share the radio spectrum equitably.

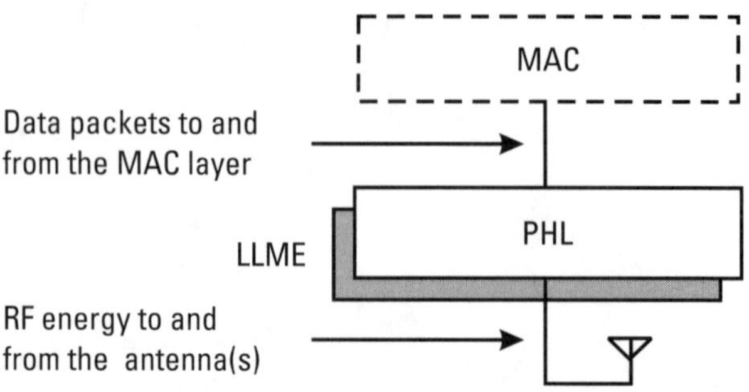

Figure 4.1 The position of the PHL relative to the other layers.

This chapter describes the DECT and PWT radio interfaces, their key characteristics, and some of their more important consequences for real equipment.

4.2 Radio Frequency Access

The radio frequency bands for DECT and PWT are illustrated in Figure 4.2.

The main difference between DECT and PWT lies in the arrangement of their radio frequencies. Most of the differences, therefore, are described in this chapter. (There are, however, several more consequences of this different radio channel arrangement at higher protocol layers, principally the MAC layer, which will be covered later.)

4.2.1 DECT Basic Frequency Band

The basic DECT frequency band has been harmonized across Europe, and European law requires that every country in Europe make those frequencies available for use by DECT systems. The common frequency band creates the possibility of a unified market for cordless products. All DECT applications—residential, public, and office—share the same frequency allocation.

DECT has a total spectrum allocation of 20 MHz, from 1880 MHz to 1900 MHz. That allocation is divided into 10 carriers, each separated by 1.728 MHz. Unlike previous generations of cordless equipment, in which each cordless set used one randomly chosen frequency from a pool of permitted frequencies, all DECT equipment is capable of working on any DECT frequency. That imposes a small cost penalty but ensures that the cordless equipment can always choose the best possible channel and that a high-quality link is maintained even in the presence of strong co-channel interferers, including other DECT system users.

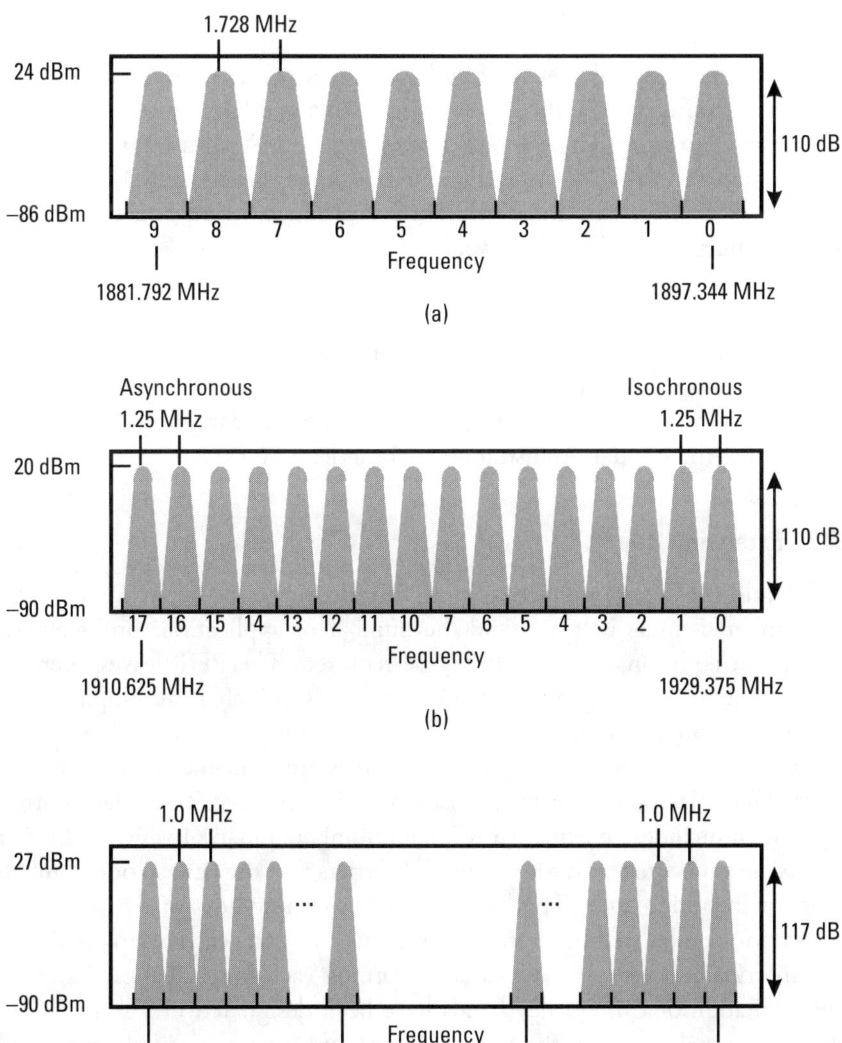

Figure 4.2 (a) DECT, (b) PWT, and (c) PWT-E spectrum bands.

The center frequency of a DECT radio frequency carrier, *Fc*, is given by this expression:

$$Fc = 1897.344 - c \cdot 1.728 \text{ MHz}$$

$$\text{where } c = 0 \ldots 9$$

The radio signals are transmitted in bursts. There is, therefore, a difference between the peak power value and the average power value, which explains a common confusion over the power values. When a DECT system transmits a burst, it may do so with a peak power of up to 250 mW. If it transmits during one timeslot out of the 24, the average RF power is approximately 10 mW. The peak power figure is the more relevant for product design and propagation planning purposes, while the average value is useful for assessing battery lifetime.

DECT has two permitted output power levels: the standard level of 250 mW and a low power mode in which the peak power is 2.5 mW. The basic standard allows manufacturers the freedom to choose their own power levels up to those peak levels, although the interoperation standard for voice systems requires that the standard transmit power be at least 80 mW.

4.2.2 Extension Bands

DECT and PWT are intended as multipurpose, future-proof technologies. As the number of users increases and the number of applications proliferate, at some point additional spectrum may be required. The PHL leaves open the possibility of adding additional frequency bands, which is accomplished by base stations broadcasting on their beacon transmissions (see section 5.5) an indication of whether any extra radio channels are available. If there are any additional carriers, the base station also transmits extended RF carrier information, which includes an extended RF band number, a map of which individual carriers are supported, and the number of carriers that the base station regularly scans for incoming calls. The handset needs to know how many carriers the base station regularly scans, even if it cannot use the extra carriers, to predict the time interval between the base station scanning each individual carrier. As of 1998, no additional frequency bands have been designated in Europe. If and when they are designated, the band numbers and their carrier frequencies will be specified in a new edition of the PHL standard.

4.2.3 PWT and PWT-E

PWT operates in two types of frequency spectrum in the United States, licensed (Figure 4.2(b)) and unlicensed (Figure 4.2(c)). Furthermore, within the unlicensed band, two types of operation are distinguished: asynchronous and isochronous. Hence, there actually are three types of PWT.

The term *licensed* refers to the fact that to operate in this band, the FCC (or alternative regulatory bodies in other countries operating PWT) must give an operator a license to use the spectrum before the operator may provide a service. The PWT-E system (sometimes called DCT1900) is intended to operate in licensed spectrum.

The term *unlicensed* means that any member of the general public may use equipment in the frequency band providing that the equipment obeys all the rules for sharing the band fairly. PWT operates in unlicensed spectrum, in both the asynchronous and isochronous subbands. Regulations for operating unlicensed equipment are found in the FCC rules part 15, subpart D, "Unlicensed Personal Communications Service Devices" [4], often called the UPCS etiquette.

The licensed band rules are different from the unlicensed band rules because licensed operators have exclusive control over their own blocks of spectrum and can manage interference by planning where transmitter sites are located. In the shared unlicensed band, it is not possible to rely on any such planning or even coordination among users, and the equipment itself must as far as possible automatically coordinate between adjacent systems. One important difference is in RF power output. Generically, PWT allows four power classes: low, standard, intermediate, and high (Table 4.1).

Handsets in the unlicensed bands may use standard or low-power modes. Handsets and base stations in the licensed bands also may use the intermediate or high-power modes.

These additional power classes offer the prospect of increased ranges (with the high-power modes) and increased capacity (with the low-power modes). The additional high-power mode for licensed base stations allows larger cells and should allow the operator to reduce significantly the number of cell sites and thus the investment needed to cover a given area. It also allows for planning of balanced links, since it often is possible to have better receive performance at the base station than at the handset.

Another significant difference is the different channel spacings in the licensed and unlicensed bands. The unlicensed bands use a carrier spacing of 1.25 MHz, and the licensed bands use a spacing of 1 MHz. Carriers may be

Table 4.1
PWT Power Classes

Power Level	Power Class	Power (mW)
1	Low	2
2	Standard	90
3	Intermediate	200
4	High	500

squeezed closer together in the licensed bands because the operator of a system has the ability to control any self-inflicted interference.

All unlicensed PWT equipment has to be capable of operating in the isochronous subband. If the equipment also can comply with the asynchronous etiquette, then it has the additional option of also using the asynchronous subband. PWT base stations broadcast a list of which mandatory isochronous carriers are available on the base station and an indication of whether any extra carriers are available. The extended RF carrier information message indicates which channels, if any, are available in the asynchronous band and which, if any, of the licensed band channels are available. Hence, the asynchronous subband actually is an extension band for PWT, with slightly different rules. At the time of writing (1998), some suggestions have been made to extend the FCC rules to permit the operation of devices complying only with the isochronous etiquette in the asynchronous subband. Readers should consult the latest rules from the FCC.

The center frequency for the PWT and PWT-E radio frequency carrier, *Fc*, is given in Table 4.2.

The FCC rules also permit unlicensed operation according to the asynchronous etiquette in the 2390 to 2400 MHz band, although the PWT standard does not define carriers in that additional spectrum.

4.2.4 Unlicensed PWT Etiquette

The FCC's spectrum-sharing etiquette [4] requires the PWT system to detect other non-PWT interferers that may have very short pulsed transmissions. In the isochronous band, the received signal strength indicator must be able to detect pulses of 50 μs when they are at a specified sensitivity threshold of 30 dB

Table 4.2
PWT Frequency Bands

Frequency Range (MHz)	Type of Operation Permitted	Number of Carriers	Center Frequencies
1850 to 1910	Licensed PCS	12 bands of 5	$Fc = 1{,}934.5 - 5n - c \cdot 1.0$ MHz ($n = 1, \ldots, 12$, $c = 20, \ldots, 24$)
1910 to 1920	Unlicensed asynchronous	8	$Fc = 1931.875 - c \cdot 1.25$ MHz ($c = 10, \ldots, 17$)
1920 to 1930	Unlicensed isochronous	8	$Fc = 1929.375 - c \cdot 1.25$ MHz ($c = 0 \ldots 7$)
1930 to 1990	Licensed PCS	12 bands of 5	$Fc = 2019.5 - 5n - c \cdot 1.0$ MHz ($n = 1, \ldots, 12$, $c = 25, \ldots, 29$)

above thermal noise power for the occupied bandwidth and 35 μs when 6 dB above that level.

The FCC rules for the unlicensed bands also limit the gain of PWT antennas to 3 dBi. In the licensed bands, PWT-E operators may use a larger antenna gain to suit their link budgets and coverage requirements. The DECT operator may use antenna gain values of 12 dBi or even 22 dBi with special permission.

A more complete outline of the asynchronous and isochronous etiquettes and the FCC rules can be found in [5].

4.2.5 Modulation

DECT and PWT use different modulation schemes. The differences arise because DECT was specified several years before PWT. The DECT committee elected for a relatively simple scheme and planned to achieve very high overall spectrum efficiency through the use of small cells. By the time PWT was standardized, there was simultaneously more experience with the production of more complex modulators in mass market quantities and more pressure for the efficient use of radio spectrum. Moreover, the PWT channel spacing had to be adapted to the U.S. bands, which presented the opportunity to use an alternative modulation scheme.

DECT uses a 1-bit-per-symbol modulation, Gaussian frequency shift keying (GFSK). It has a bandwidth-time (BT) product of 0.5, making it equivalent to Gaussian minimum shift keying (GMSK). GMSK has a relatively narrow transmitted bandwidth with constant amplitude and can use simple, inexpensive, power-efficient, nonlinear transmitter amplifier circuits. GMSK also allows simple receivers with noncoherent reception, bit by bit decision, and easily implemented IF filters, and it may be noncoherently demodulated using a simple limiter discriminator. The digital data can be recovered using clock recovery and bit-slicing circuits. The clock recovery circuit phase locks to the symbol timing of the incoming data and produces a pulse at the center of the bit period. The bit slicer is a comparator that recovers the modulated data by setting a reference level midway between the binary states and comparing the received signal with that reference.

PWT uses a 2-bits-per-symbol modulation, $\pi/4$ differentially encoded quadrature phase shift keying ($\pi/4$ DQPSK). That scheme is more spectrally efficient in that it squeezes the same data rate into a narrower bandwidth. In quadrature phase shift keying (QPSK), the signal may take one of four different phases (say, 0°, 90°, 180°, 270°), and each phase is taken to represent 2 bits of data. So, where every GMSK symbol transmitted represents only 1 bit of information, each QPSK symbol represents 2 bits. It is that factor that allows the

1,152-kbps data rate to be squeezed into a smaller transmit bandwidth than in DECT. QPSK can be thought of as two independent channels each modulating the (orthogonal) in-phase and quadrature components of the carrier. With QPSK, when the phase changes from 0° to 180°, the signal amplitude must pass through zero, so there is an amplitude component in the transmitted signal. The amplitude is constant at the sampling intervals but varies during the phase transitions. That amplitude variation requires the use of linear power amplifiers in transmitters. If the amplifier is not linear, the amplitude variation is distorted and the transmitted spectrum increased.

The $\pi/4$ shifted DQPSK is intended to reduce the amount of amplitude variation compared to QPSK, though it still requires the use of linear amplifiers. It restricts the maximum amplitude change by restricting the maximum phase change. Although it is a quaternary modulation scheme, there are, in fact, eight allowed phase states, each separated by $\pi/4$. The eight phase states are organized as two sets of four phase states offset by $\pi/4$, that is, 45°. The modulation rules limit the maximum phase change during a single symbol period to either $\pi/4$ or $3\pi/4$, that is, less than permitted in QPSK, which limits the amplitude variation. The scheme thus requires subsequent symbols to use phase symbols from a different set of four phases. That makes symbol clock recovery easier because there are always phase transitions at the symbol rate. $\pi/4$ DQPSK uses differential encoding whereby the information to be transmitted is encoded in the phase difference between subsequent symbols rather than the absolute value of the phase relative to a reference. That means receivers can be simpler because they do not have to recover the exact phase of the reference carrier but may demodulate by comparing the current phase with that of the previous symbol.

Normally, a modulation scheme with more phase states has a higher probability of error since it takes a smaller noise signal to make one state appear as another. Although $\pi/4$ DQPSK is a four-state scheme, it has the same error performance as a two-state scheme. That is because the quadrature modulation means that the in-phase and quadrature channels are, in effect, independent two-state modulation schemes.

4.2.6 Frequency and Timing Control

DECT and PWT do not call for stringent accuracy of the RF carrier center frequency or of timing jitter, which helps keep equipment costs down. The standards describe mechanisms for providing higher accuracy frequency and time references at the base station and using those mechanisms to compensate for inaccuracies or offsets in the handsets. The accuracy at the base station is ±50 kHz, and the handset may either lock to the received carrier or use an independent crystal as long as it provides the same level of accuracy. The standards do not provide any special frequency- or time-correction signals; again, that is a trade-off between maximizing performance and maintaining a reasonably simple affordable system. Instead, a closed-loop feedback mechanism is provided for both frequency and timing and is operated by the MAC layer. The base station measures any error in the frequency received from the portable and then informs the portable of the measured error (using a MAC-layer message called *quality control*). The portable then can adjust its frequency or, if it does not have ability to fine-tune its carrier frequency, ignore the fixed part's advice.

Similarly, the base station may measure any timing offset and can instruct the handset to adjust its transmit timing. The timing adjustment feature is valuable in large cell systems, in which the time of flight or propagation delay from the handset to the base station is similar to the intertimeslot guard time. For example, if a base station with a directional antenna serves a handset at a distance of 3,000m, then the time of flight to the handset is 10 μs and the return flight of the signal from the handset takes another 10 μs, giving a total time of 20 μs. The total guard space is about 30 μs (a total of 50 μs less 10 μs for the transmitter to reach full power output from idle and 10 μs for power-down ramping). If the extended preamble also is used, the guard space is further reduced. If the mobile is too far away or the guard space is reduced, perhaps by a timing offset at the portable, then at the base station the reception of the end of one user's transmission may overlap with the beginning of another user's transmission. The timing correction facility prevents that overlap.

4.3 Multicarrier Time Division Multiple Access

Many of DECT/PWT's valuable characteristics arise because of its highly flexible multicarrier TDMA structure and the way in which it defines its physical channels. It is flexible for two reasons: first, compared to other radio systems, it has a very broadband transmission rate, and second, there are many different permutations of its transmissions. It is instructive to think of a 24-slot system with a 10-ms frame in which it is possible to use any slots in more or less any combination. Normal telephone calls use slot pairs separated by 12 slots (half a frame).

4.3.1 Frame and Slot Structure

Information is transmitted on each DECT or PWT carrier at a rate of 1,152 kbps. That is more or less the highest bit rate practical and economical for a short-range indoor radio system. As illustrated in Figure 4.3, each radio

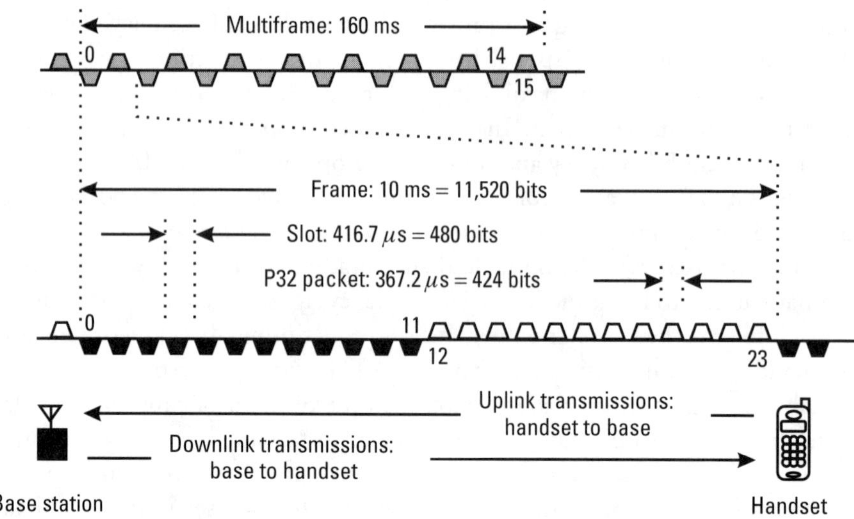

Figure 4.3 Frame structure and duplexing.

carrier is divided into regular cycles of 24 timeslots. One complete cycle lasts for 10 ms (11,520 bits) and is called a frame.

Normally, during the first 12 timeslots of the frame, the base station transmits to the handset; during the second set of 12 timeslots, the handset transmits to the base station. Each transmit slot has a corresponding receive slot, 12 slots later. The process is called time-division duplexing because it allows both handset and base station to communicate with each other, apparently simultaneously, using only a single carrier rather than separate carriers for transmitting and receiving.

Figure 4.3 also illustrates how several frames are grouped together to form a multiframe. The multiframe, a concept introduced by the MAC layer, is the period over which a complete set of system information is broadcast, and is thus the period over which the handset has to listen to a base station's beacon to get a complete picture of the system to which it is listening. (Multiframes are explained in more detail in Chapter 5.)

To transmit a (continuous) speech signal using a regular series of transmission bursts, the speech signal must be buffered, as illustrated in Figure 4.4. The continuous analog speech signal is digitized, encoded at 32 kbps and read into a buffer at that rate. Over the period of one frame (10 ms), 320 bits of speech data are accumulated. When it is time to transmit, the data are transmitted out at a much higher rate (1,152 kbps), and the entire contents that took 10 ms to record are transmitted in one slot during less than 416.7 μs. It is

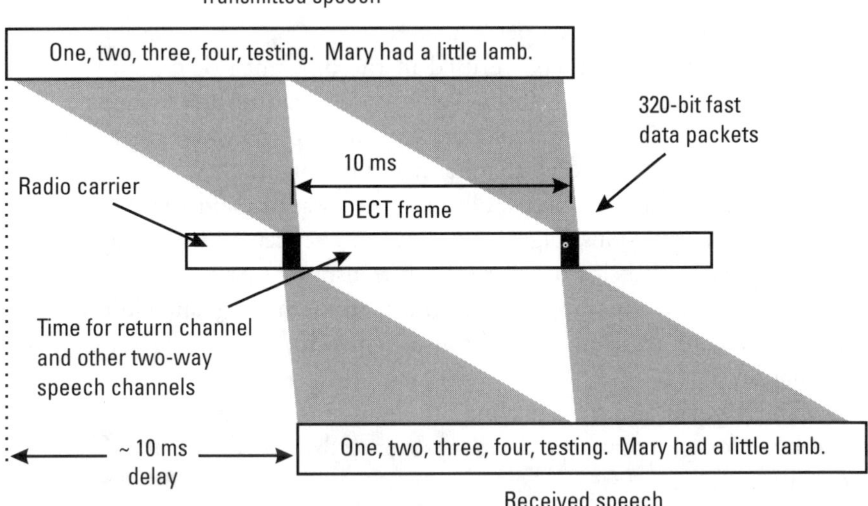

Figure 4.4 Speech buffering.

that compression in time, by a factor of more than 24 that allows 24 timeslots, 12 transmits and 12 receives, to be multiplexed onto a single radio carrier.

Notice that the buffering process introduces a pronounced delay in the speech transmission. Delay in speech may be undesirable and may be objectionable if telephone users hear delayed acoustic echoes of their own voices. System installations typically introduce echo-control measures to ensure that does not happen.

To allow data to be multiplexed into and demultiplexed from the frame, each slot and bit are numbered. The frame is divided into 24 full slots, numbered $K = 0$ to 23. The full slots are 480 bit periods long, and the bit periods are numbered f0 to f479. Each full slot may be divided into two half slots, each of 240 bits and numbered $L = 0$ for the earlier and $L = 1$ for the later half slot. Two full slots may be combined to form a double slot of 960 bit periods, numbered f0 to f959. A double slot may start only on an even-numbered full slot and is designated using that even number.

4.3.2 Packets Within Slots

Each segment of speech data is transmitted within a data burst called a *packet*. The word *packet* is used to distinguish between the actual transmission and the timeslot. Timeslots are intervals during which packets can be transmitted.

Continuous channels are created by transmitting a series of packets at regular intervals, normally one packet per frame.

Different-length packets are used in the different slot types to provide channels with different bandwidths. Table 4.3 summarizes and Figure 4.5 illustrates the types of channel available, the slots in which they are transmitted, the packet used, and for what each channel typically is used.

To transmit speech successfully, it is necessary to add various signaling and control transmissions. Figure 4.6 gives an overview of one of the packets within the frame and slot structure. Each transmitted packet comprises what we shall term, for the moment, "start" information, then signaling information, then payload information (e.g., the user's speech), and finally some "stop" information.

The payload and the signaling are the useful information, and the start and stop information are the necessary overheads needed to make the system work successfully.

The start information comprises some preamble, which serves to let the receiver warm up before receiving real data, followed by the synchronization marker. This field marks the beginning of a packet and in a time division system is a critical function. The fact that there is a synchronization marker in each packet simplifies equipment design by letting the handset and the base synchronize to each other in each packet rather than having to maintain highly accurate clocks in both the handset and the base station. In particular, it simplifies the handset receiver, in which the regular synchronization marker allows the handset to lock to the base station and avoids the need for design of stabilized clock systems. Optionally, there can be two preamble fields, which can be used to assess if the beginning of the packet has been corrupted by transmissions from adjacent slots from other nearby base stations (discussed in Section 4.4.5), in which case the outer field will be corrupted regularly, but the

Table 4.3
Slot and Packet Types

Channel	Slot	Packet	Traffic Rate	Typical Use
Short	Full	P00	0 kbps	Beacon
Low rate	Half	P08j	8 + *j* kbps	"Half-rate" speech codec
Basic	Full	P32	32 kbps	Normal (32 kbps) speech
High capacity	Double	P80	80 kbps	64 kbps PCM and ISDN

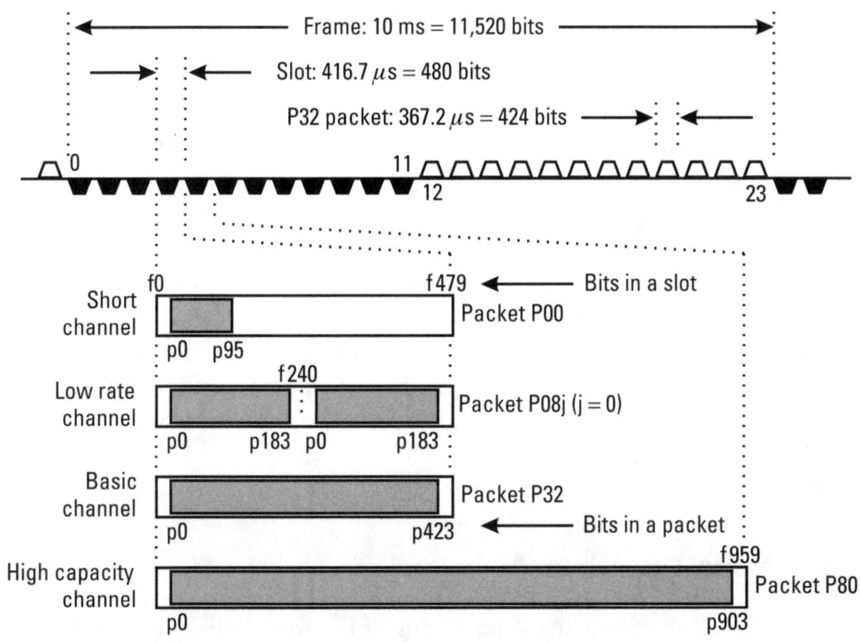

Figure 4.5 Types of slots and channels.

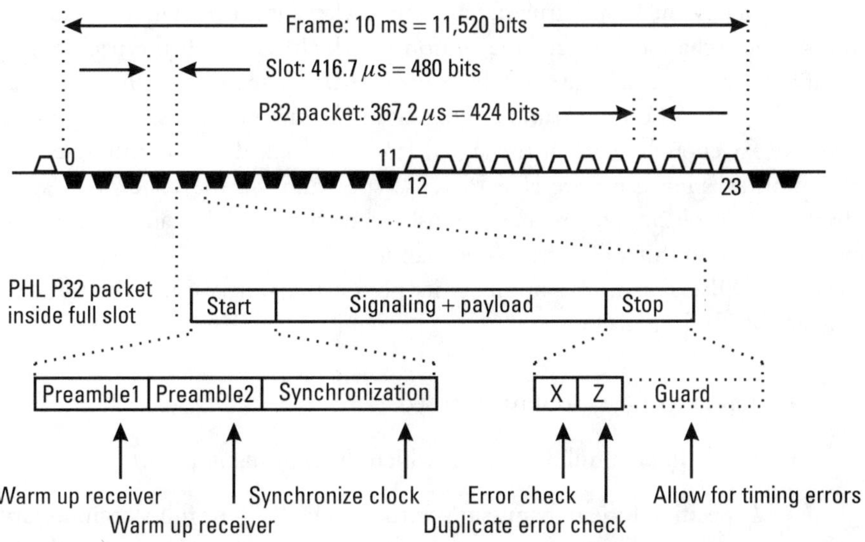

Figure 4.6 A packet within the frame and slot structure.

inner field will mostly be correct. It also facilitates antenna diversity (discussed in Section 4.5.4).

The stop information comprises some error-check data and some guard bits. The error-check data are intended as a means of monitoring the quality of the channel, perhaps as a prelude to taking a decision to change antenna or channel, rather than as a means of detecting and correcting errors in the speech channel. Notice that the error-check field optionally may be duplicated.

For clarity, the formal protocol layers have been somewhat mixed in the preceding description. The data packet that comes from the MAC layer comprises the signaling, the payload, and the X-field error check. The PHL calls the complete packet the D-field. The PHL is responsible for duplicating the X-field and calls that the Z-field. It also adds the preamble and synchronization fields, which together it calls the S-field.

4.3.3 Physical Channels

Technically, all the foregoing descriptions of the frame structure and the data packets are needed to split the assigned radio spectrum into physical channels. The reason is that we need to be able to give the MAC layer some means to ask for the radio resources it wants to use, and the physical channel is the concept that the MAC uses in communicating with the PHL.

As well as the frame structure and the packets, it is necessary to include in the physical channel concept the notion of which radio cell the packet uses. That is because over a large enough area covered by many radio cells, more than one radio cell can use the same frequency in the same timeslot, as long as the cells are far enough apart so the signal from one link does not interfere with the another. Hence we are able to reuse spectrum via the concept of space division multiplexing, or frequency reuse, whereby two links can use what you normally would think of as the same channel.

The DECT standard numbers its physical channels using the notation Ra(K,L,M,N). In that notation,

- *a* specifies the type of packet used.
- *K* specifies the full timeslot in which the transmission starts.
- *L* specifies if the transmission starts on a half-slot or full-slot boundary.
- *M* specifies which radio carrier is used.
- *N* specifies which cell is used (the RFP number).

A few qualifying parameters for a physical channel indicate whether some of the optional fields are used:

- *s* specifies if the extended preamble is used.
- *z* specifies if the Z-field error check is available.

The short physical channel (see Table 4.3) is used to transmit a beacon signal or "dummy bearer" from a base station from which there are no other traffic channel transmissions. It uses a short packet P00 (96 bits) in a full-length slot. Formally, it is termed the R00 channel, which indicates that it is a zero-rate channel that carries no traffic. The terminology R00(K,L,M,N) is used to fully designate a particular instance of this channel.

The basic physical channel (R32) is used to carry normal 32-kbps ADPCM-coded voice transmissions. It uses a basic physical packet P32 in a full slot.

The low-capacity physical channel (R08j), or "half-rate" channel, is intended to carry traffic from an anticipated "half-rate" speech codec. It uses a P08j packet in a half slot. The terminology R08j indicates that the channel has capacity for a data rate of 8 kbps of traffic and that there is the potential to increase the user's data rate by some amount *j*, to 8 + *j* kbps. In practice, increasing the data rate of the channel by increasing the number of bits carried in each half slot would reduce the guard space between adjacent half slots.

The high-capacity physical channel (R80) can be used to carry high data rate services such as conventional telephony using full PCM at 64 kbps, including ISDN B-channels. In fact, it has capacity for up to 80 kbps, using the P80 packet in a double slot.

4.3.4 Packet Structure

Figure 4.7 illustrates the internal structure of a packet. Strictly, this is a matter for the MAC layer, and the PHL does not need to know anything about what is in the packets it sends and receives. The MAC layer defines the PHL's data field in each packet, the D-field, to contain an A-field (64 bits), a B-field (whose length depends on the type of packet), and an X-field (4 bits).

The A-field contains signaling within several logical channels that are multiplexed together. It is organized into the A-field header (8 bits) and the A-field tail (40 bits). The A-field header is used to identify the type of logical channel in the A-field tail and to identify the type of traffic in the B-field. The header also is used to transmit the information that is used to monitor channel quality and may be used for deciding if a new physical channel should be selected. The A-field tail contains the logical signaling channels themselves,

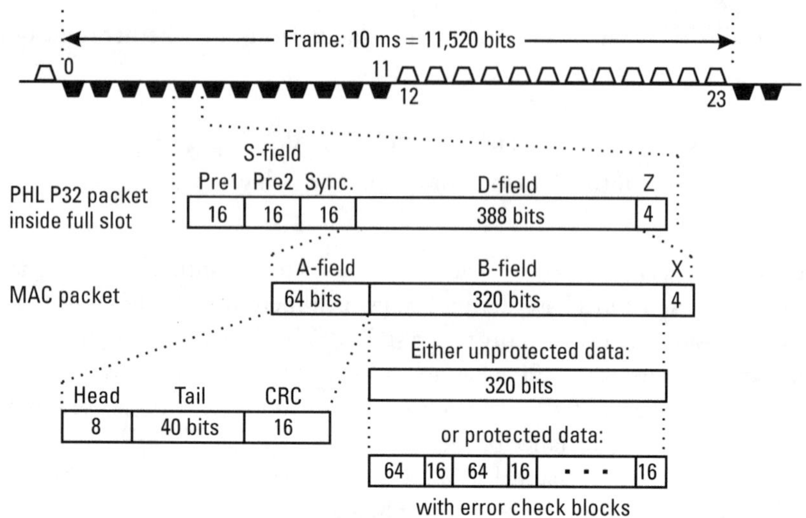

Figure 4.7 Packet structure.

such as system identity information, system operating information (e.g., the number of transceivers at that base station), handset paging information, or internal MAC-layer signaling. The 64-bit A-field also contains a 16-bit CRC (cyclic redundancy check) error check field, which is calculated over the remaining 48 bits.

The B-field is the payload, the field that carries user traffic. In a typical voice connection, the B-field is 320 bits long, which, with 100 frames every second, yields a throughput data rate of 32 kbps. The B-field may employ either protected or unprotected modes. The unprotected mode does not add any additional coding and thus has a maximum throughput of 32 kbps. The protected mode divides the B-field bits into groups of 80 bits, each comprising 64 bits of data and a 16-bit CRC. A higher layer application has the choice of requesting a protected mode or of using the unprotected mode and adding its own error correction protocol if needed.

The S-field comprises a preamble or a pair of preambles and the synchronization field.

4.4 Physical Layer Operation

The structures the PHL defines are the basis for its operation. This section covers some of the key operational issues, including channel allocation, data transfer, and handover.

4.4.1 Dynamic Channel Allocation

In choosing a physical channel on which to operate, the system uses dynamic channel allocation, or DCA, which is based on two principles: a portable choosing to use the base station with the strongest received beacon signal and both the handset and the base station keeping a record of the received signal strengths on physical channels they are not currently using, to be able to select one with the least interference whenever the need arises.

When the portable is first switched on, it is not yet locked to a base station. The lock is accomplished by searching all channels for base stations broadcasting system identities that the portable recognizes and then selecting the one with the strongest signal strength. When the portable has locked to a base station, it continues to monitor all other channels (or as many as it can) to determine if there is another suitable base station with a better signal quality. If a better base station is found, the portable may relock to that one.

When establishing a call, the portable and the base station choose the best mutually acceptable channel based on their channel signal strength maps. During the call, the portable continues to monitor the signal strength on adjacent base stations and alternative physical channels. It may decide to hand over the call to a better channel or better base station if one can be found.

The systems thus are constantly measuring the signal strength and signal quality of many channels. The standard formally describes the process by saying that the LLME instructs the PHL to measure a particular channel by sending a PL_ME_SIG_STR-req primitive to the PHL. The PHL then measures the signal strength and returns the value to the LLME in a PL_ME_SIG_STR-cfm primitive. The received signal strength indicator (RSSI) measurements passed in the PL_ME_SIG_STR-cfm primitive may be quoted in "internal" units rather than in real units (of watts or dBm). In other words, the signal strength can be measured in arbitrary units, which then can be calibrated to correspond to real units. The LLME issues a series of measurement commands to the PHL and then produces an ordered list of the base stations with higher signal strengths (wanted signal) and potential traffic channels with lower (unwanted interference) signal strength. That procedure allows the portable to maximize the C/I when selecting a new channel.

Transmission bursts often are very short. A basic packet P32 has a duration of only 424 bits or 368 μs, and a dummy bearer beacon transmission, on a P00 short physical packet, has a duration of only 96 bits or 83 μs. The RSSI measuring circuit therefore must have quite a short time constant if it is to detect and respond to them. When measuring channels, the receiver must measure the signal over the entire length of the packet transmission plus 10 μs before and after with a circuit with a time constant of between 10 μs and 40 μs.

The PWT etiquettes require further specific performance for detecting short transmissions, since systems other than PWT may be operating in the same radio band.

The portable is required to measure the received signal strength with a resolution of better than 6 dB, that is, if two signals differ by more than 6 dB, the portable can tell them apart. It also is required to maintain that resolution over a range of signal levels between –33 dBm (a strong signal) and –93 dBm (a very weak signal). That is not trivial to achieve because many components used in real products are intended to measure relatively rather than accurately. In many products, the received signal strength is measured by taking an RSSI output from a demodulator integrated circuit and connecting that to the analogue-to-digital-converter input on a microprocessor. Often the RSSI circuit is quite simple and intended only to provide a linear output over only the low end of the range, because many applications such as opening a squelch or finding a free channel are interested only in weak signal values. Some circuits may even have a multivalued output, with the RSSI output falling if the signal strength increases too far. The standard insists that signals less than –93 dBm and more than –33 dBm should produce outputs that are less and greater, respectively, than the outputs produced by those values.

As well as measuring the signal strength of a candidate base station, the portable also may measure the "quality" of the signal. The quality of a signal is important for interference-limited systems, but it is not easy to measure. The best measure of channel quality is the C/I, but that cannot be measured directly because the wanted and unwanted signals are indivisibly combined. In practice, it is necessary to measure some analog of the C/I. It is common to use some signal from the demodulator, for example, a signal representing the eye opening or a measure of the amount of jitter on the recovered clock.

4.4.2 Data Transfer

The process of transmitting a packet is described by a procedure in which the MAC layer supplies the PHL with a PL_TX-req primitive containing the data packet to be transmitted, the physical channel on which the packet is to be transmitted, and an instruction on whether or not to add a Z-field.

Similarly, the process of receiving a correctly synchronized packet is described by saying that the MAC layer instructs the PHL, with a PL_RX-req primitive, to receive a specified packet of data from the specified channel. The PHL layer then acquires synchronization, receives the packet, strips away the synchronization word and the Z-field if it is present, and passes the data to the MAC layer in a PL_RX-cfm primitive. The PHL advises the MAC layer if the synchronization word was correctly received, supplies the D-field, and may advise the MAC if it has measured a frequency error or a sliding collision (described in Section 4.4.5). The standard also describes the process of searching for a synchronization word on a new channel by saying that the LLME sends the PHL a PL_ME_SYNC-req primitive.

4.4.3 Handover

The broadband TDMA structure allows a powerful form of seamless handover, a make-before-break technique in which a connection is established on a new channel before the old one is dropped (Figure 4.8). The technique is possible with little penalty because the TDMA structure allows two channels to be established to the same portable. There is very little extra complexity at the portable to allow it to establish parallel channels that transmit and receive the same information.

Handover usually is initiated by the portable because it is in the best position to assess interference conditions. There are various types of intracell and intercell handover at several layers of the protocol stack, (those at the MAC layer are described in Chapter 5; those at the DLC layer are covered in Chapter 6). This section discusses some of the PHL aspects of the seamless handover mechanisms.

Seamless handover is similar to what CDMA systems term "soft" handover. In a TDMA system, the portable receives the same data in two different

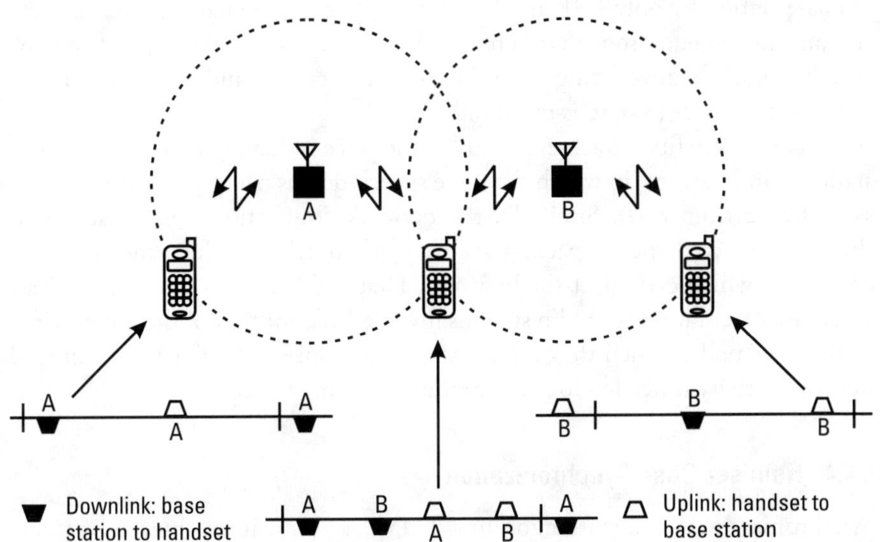

Figure 4.8 Seamless handover (macrodiversity).

timeslots, while in CDMA the portable receives the same data as two different codes received by two independent correlator branches. Soft and seamless handovers have several important advantages over hard handovers.

First, there is no perceptible interruption of the user's speech when a call is switched from one channel to another. For data connections, it may be possible to avoid the need for flow control algorithms synchronized to the handovers. By allowing a portable in an area of marginal coverage to remain connected to two base stations simultaneously, reliable coverage can be achieved even with weak radio signals at the edge of a cell.

Compared to hard handover, seamless handover has the effect of increasing cell sizes. It allows a base station controller to implement a form of macrodiversity in which it listens to signals from the same portable received via two different base stations and on a packet-by-packet basis choose the packet with the best received quality. In an attempt to prevent handovers from becoming too frequent in hard handover systems, there usually is some hysteresis built into the algorithm. The new base station must have a signal strength typically 6 to 8 dB greater than the old base station. Thus, in a hard handover system, if the old base station has a signal strength of n dB and the prospective new base station has a signal strength of $n + 5.9$ dB, the link will not be able to benefit from that potentially better signal. Moreover, the new base station normally is required to be greater than the old signal for a specified period, further reducing the potential to exploit the fact that the new signal is better than the old signal. In a seamless handover system then, if the new base station is better than the old base station by only 1 dB and for only one packet, the handset and the base station may benefit from that. The magnitude of the benefit depends on the correlation of shadow fading from the two base stations and the magnitude of the hard handover hysteresis threshold.

Those benefits are achieved only at the price of having synchronized base station transmissions, in which the same speech data is being transmitted in the same frame from each. Similarly, the network connection must ensure that the same received speech packets are brought together at the same time to a combiner, which can select the best one. That is difficult to achieve, not least when the base stations or radio systems involved are not part of the same switch or network node. When they are suitably synchronized, then a high degree of network activity is needed to perform the selection.

4.4.4 Handset-Base Synchronization

Synchronization is a key function in a TDMA system. It is achieved by transmitting a synchronization marker with each packet of data, which is then detected at the receiver. The portable and the base station are not required to synchronize with other systems and users; through relative clock drift, that can lead to interference when transmissions in one timeslot drift into those of other users (covered in Section 4.4.5). Base stations and portables use different synchronization markers to prevent portables from locking to the transmissions of other portables.

The timing stability requirements are different for portable and base stations and different for single channel and multichannel base stations. The portable must keep its reference clock to better than 25 ppm at the extremes of the temperature ranges, which can be achieved with an ordinary crystal. Since the portable locks to the base station, the stability requirements are higher for the base station, which must achieve 10 ppm at the extremes of the temperature ranges. As well as the synchronization timing mechanisms, there also are timing compensation mechanisms that are used to compensate for the effect of long propagation delays (see Section 4.2.6).

To obtain synchronization with each other (or with neighboring systems), the portable and the base station are allowed either to adjust their relative timing in one single step or to gradually alter the length of time between successive transmissions until synchronization is achieved. To do that, the length of successive frames may be adjusted by up to two bit periods. The former method is suitable for portables locking onto a base station, the latter for a base station that is carrying traffic and that is suddenly connected to an external synchronization source.

It is possible to use different techniques for acquiring and maintaining packet synchronization. For example, you can use a different window size (i.e., the interval over which the channel is searched), a different pulse height, and different clock recovery loop time constants. When attempting to acquire synchronization, it may be necessary to search over a wide interval of time. Doing so, however, increases the probability of false detections. The synchronization word is 16 bits long. Thus, 1 in 65,536 random combinations of bits may emulate the synchronization word. Shortening the search window reduces that possibility. The GAP requires that the search window be at least 4 bits either side of the expected response if the portable is already locked to a base station and otherwise at least 10 bits on either side. While acquiring synchronization, the loop may have a fast time constant to allow it to lock on quickly. When synchronization has been acquired, the loop bandwidth can be reduced to prevent drift.

The process of synchronization is described in terms of primitives exchanged between the PHL, the MAC layer, and the LLME. The LLME orders the PHL to gain synchronization by sending a PL_ME_SYNC-req primitive, and the PHL then responds with a PL_ME_SYNC-cfm primitive when it detects a synchronization word.

4.4.5 Sliding Collisions

Whenever intersystem synchronization is not perfect, there is the possibility that nominally successive packets from adjacent systems will use up the guard space between them and collide. That is illustrated in Figure 4.9, where two users are using immediately adjacent timeslots on nearby base stations. Since the two separate base stations are not synchronized, their clocks drift slowly apart, which leads to the adjacent packets colliding. Being near each other, the two links interfere seriously with each other. However, mechanisms are provided to detect such clashes at both the beginning and the end of a packet. The portable and the base station then may perform a handover to a new channel before the user data are corrupted.

It is straightforward to detect a clash at the beginning of a packet. The packet begins with a synchronization word that has a known bit pattern. If the synchronization word had been received successfully for many packets but is suddenly received incorrectly, that may indicate a clash with a packet from another transmitter. There are several possible factors that might corrupt the synchronization word, not just a sliding collision, and the receiver may improve

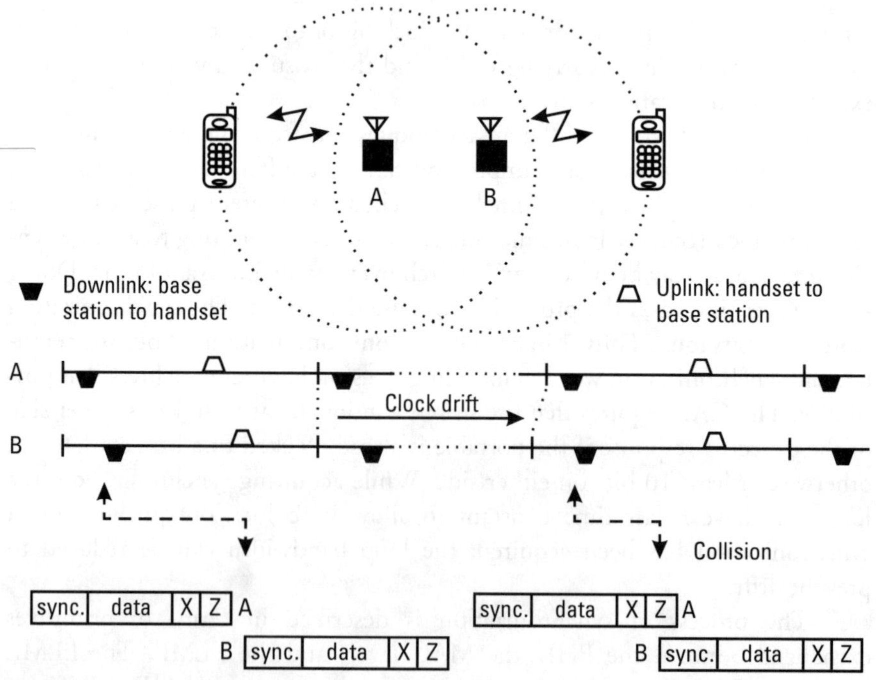

Figure 4.9 Detection of a sliding collision.

the quality of its judgment by measuring the signal strength in the period before the packet is received and by checking if only the first bits in the synchronization word are corrupted.

It is more difficult to detect errors indicating a sliding collision at the end of a packet since the packet concludes with a CRC check (the X-field) and the value of those bits cannot be determined in advance. Detected errors in the CRC field would appear to tell the receiver that there were errors in the data field. Therefore, it is possible to optionally add a duplicate CRC field, called the Z-field. If the primary CRC (the X-field) is correct, but the duplicate (the Z-field) is corrupted, that is a possible indication of a sliding collision at the end of the packet.

The base station can report any collisions it detects to the portable. Having detected or received notification of a sliding collision, the portable or base station can take the appropriate action and either attempt to adjust its own timing or hand over to a new channel.

The sliding collision mechanisms are optional in the base standards but are required by the GAP for interoperability.

4.4.6 Intersystem Synchronization

One of the requirements of TDMA relative to FDMA is that whereas frequency is an absolute concept, time is relative; that makes it more difficult to share timeslot resources among separate base stations. For example, it is much easier to allocate to one base station frequencies f_1, f_2, f_3 and to allocate another base station frequencies f_4, f_5, f_6 than it is to allocate to each base station three nonoverlapping timeslots. It is easy to use a crystal in each base station to provide a good frequency reference, but there is no natural equivalent time reference. It also is easier to give 10 uncoordinated users each a unique frequency channel than it is to give them a unique time channel. With no common timing reference, there is a danger that TDMA transmissions will partially overlap. The overlapping slots may be unusable, which will reduce the traffic capacity.

As described in Section 4.2.6, there are mechanisms to detect collisions due to clock drift and for escaping the interference by handing over to a new channel when collisions are detected. Nevertheless, there are cases in which it is necessary or beneficial to synchronize two neighboring base stations. First of all, it is necessary to synchronize two base stations to provide seamless handover between them. Then it can be beneficial to synchronize two independent but neighboring systems, especially if the base stations are close to each other, to lessen the mutual decrease in capacity due to overlapping timeslots. See, for example, [6].

When base stations are synchronized, the interference mechanisms will still be the same as those in an FDMA system. Transmissions from one base station will appear as co-channel or adjacent-channel interference to portables on the other base stations and vice versa. That interference not only will reduce the C/I of the wanted signal but also may generate intermodulation products on other channels (if the base stations are very close together).

When two base stations are not synchronized, the interference mechanisms are different. First, even if the signals only partly overlap, that ruins reception of the entire packet. Second, one base station may transmit while the other is receiving. That is conventional co-channel interference, but if the base stations are very close, the effects can be highly detrimental. The strong received signals produce strong unwanted intermodulation products. When a receiver receives two strong signals, then as the receiver's nonlinear circuits amplify those signals, a third product signal is produced. Those products are of the form $2f_1 - f_2$, which lies in the wanted frequency band and cannot be eliminated. The third interference mechanism is that as the transmitter is switched on and switched off, it produces unwanted emissions outside the channel, often called "splatter." If the transmissions are synchronized, no receivers are active during the switching period. If the receivers are not synchronized, the splatter increases the level of ambient background noise and thus reduces radio range.

There are two methods of synchronizing base stations. One is a wire-based method and the other is a wireless method based on the use of the radio timing signals provided by the Global Positioning System (GPS).

The wired method uses a cable terminated in either RJ12 modular phone jacks or A/B terminals. The base station is provided with a SYNC IN port and a SYNC OUT port using RS422 levels. The base station monitors the input port; if there is an input synchronization signal, it synchronizes itself to that and regenerates the signal at the output port. The synchronization signal is a 100-Hz signal (the 10-ms frame rate). To allow base stations to acquire multi-frame as well as frame synchronization, each sixteenth pulse is wider (between 2 ms and 5 ms) than the others (which are between 0.5 ms and 1 ms). Synchronization for the scan carrier sequence or the encryption mask must be communicated between base stations by other wired signaling means.

In cases in which several base stations are daisy-chained together, propagation delays along the cables and through the regenerating circuits may accumulate so that the base stations at the beginning and the end of the chain are synchronized, but not in phase. To prevent that, especially for cases in which it is necessary to permit handover from one base station to another, the base stations may include a timing compensation circuit.

If it is not possible to synchronize two base stations using a cable, for example, because the two base stations are on opposite sides of a public highway, then the base stations can be synchronized using GPS. A typical GPS receiver provides a "tick" every second. However, synchronization is required for the 10-ms frame and the 160-ms multiframe. The base station thus takes every fourth tick from the GPS receiver as the synchronization reference for the start of a frame and the start of a multiframe. That 4-second interval contains 25 multiframes and is called a *hyperframe*.

It also is possible to synchronize a multiframe number used by two base stations. The multiframe number is a 24-bit number that changes every multiframe and is used to synchronize encryption. The base station may obtain from the GPS receiver the time of day expressed as an integer number of seconds and scale that number by a factor 25/4 and reduce it to a 24-bit number by taking the remainder modulo 2^{24}. It is further possible to synchronize the primary carrier scan sequence, which is the sequence in which a base station's first free receiver scans channels for calls. The first frame of the hyperframe is always set to scan RF carrier 0, and the base station advertises on its beacon the number of carriers it supports. Since the base station scans at one carrier per frame, that allows the portable to know at all times where all base stations will be listening for calls. If further carriers are allocated in the future, this method continues to work.

4.5 Practicalities of Transceivers

One of the major reasons for adopting a wideband TDMA structure was the fact that it promised simpler and hence lower cost transceivers, especially for multichannel base stations. In particular, TDMA avoids the problems of transmitter and receiver intermodulation as well as the problems of adjacent channel interference.

Figure 4.10 illustrates the structure of a typical transceiver. It is drawn to illustrate block-level concepts rather than as a real circuit schematic. There are many potential variations and on the basic structure. The illustration is of a double superheterodyne system, the most common form used.

Examining Figure 4.10 from left to right, we can make the following observations. The transceiver uses a pair of diversity antennas. Normally, diversity needs to be implemented only at the base station since the channel is reciprocal over short intervals of time. The switches commonly are implemented using PIN diodes. A microprocessor may control the switching, depending on the relative strength of the signal at either antenna.

That then can be followed by a bandpass filter. On the receive side, a bandpass filter serves to filter signals from immediately adjacent bands and to remove distant frequencies that the mixer inherently but undesirably would

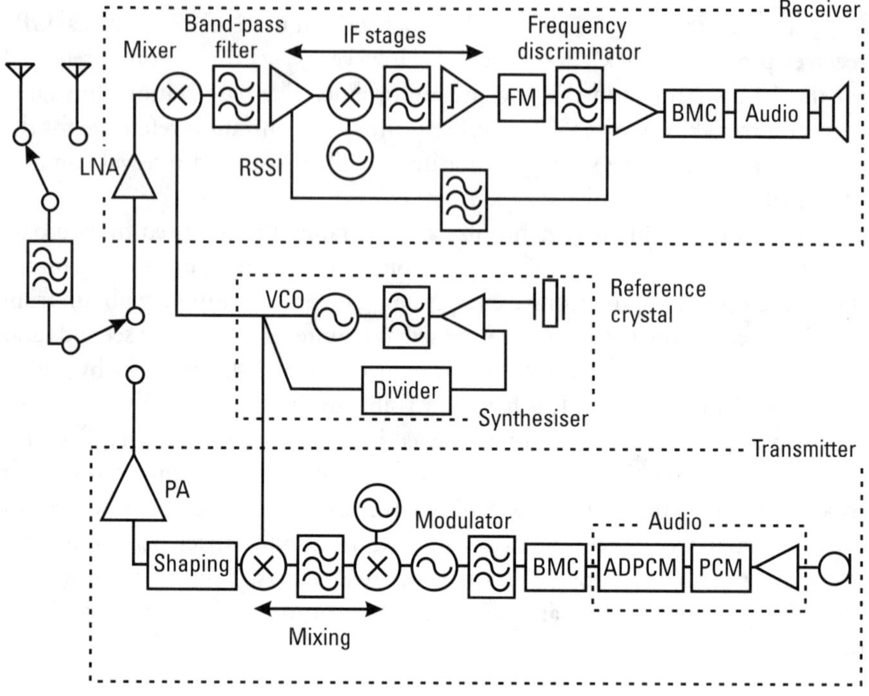

Figure 4.10 Structure of a typical DECT transceiver.

convert to the wanted frequency band. The filter may have to filter out strong signals from neighboring DCS1800 or PCS-band transmissions to prevent receiver blocking. Blocking is a mechanism whereby strong signals swamp the front end of a receiver. The receiver blocking requirements state that the receiver must withstand a signal of –43 dBm 6 MHz either side of the operating band. That region includes the DCS1800 or PCS base station transmit band, and it may be advisable to design for higher than the minimum value in the standard. On the transmit side, the bandpass filter removes unwanted emissions, including the harmonics of the fundamental frequency.

The transmit and receive sides are separated with a transmit-receive switch, which normally is implemented as a PIN diode. The important requirements of the switch are that it has a low noise figure, a low resistance during the on phase, and high isolation during the off phase. In multitransceiver implementations, the switch can be replaced by a circulator because of the circulator's higher power-handling capabilities.

4.5.1 Typical Receiver Structure

The most important requirements of the receiver are that it can detect very weak signals (its sensitivity) and that it has the ability to withstand the presence of strong unwanted signals (blocking and intermodulation). Those requirements tend to be mutually incompatible. A goal of high sensitivity calls for high gain in the receive amplifiers, while a goal of high resistance to strong signals calls for low gain. The gain of each of the amplifiers must be carefully planned to achieve acceptable performance.

The first element in the receive chain is the low-noise amplifier (LNA). The LNA is a critical element because it has a dominant effect on the received signal-to-noise ratio (S/N) and thus the sensitivity. The key parameter of the LNA is its noise figure (or noise factor in linear units). That is the amount by which the LNA worsens the received S/N; the lower it is, the better. It is critical because it is impossible to recover the S/N after it has been reduced. The noise figure of subsequent stages is not so important because it is possible to compensate for that by amplifying the signal and protecting the S/N before the stage with the poor noise figure. For that reason, many implementations introduce the receive filter after the LNA because a passive filter usually has a high loss (>3 dB). Depending on its position in the receive chain, the passive filter can substantially increase the overall noise figure and reduce the sensitivity. For that same reason, it may be desirable to increase the LNA gain as much as possible. However, increasing LNA gain too much will cause problems in the next stage.

The next stage in the receive chain is the mixer, which mixes (multiplies) the received RF signal with a locally generated oscillator signal to produce an intermediate frequency (IF) signal. All subsequent amplification and filtering can be performed at the IF. When two sinusoidal waveforms are multiplied, the result is two products, one at the sum frequency ($f_1 + f_2$) and one at the difference frequency ($f_1 - f_2$). The output frequency of the local oscillator (LO) may be either greater than the carrier frequency (high side) or lower than the carrier frequency (low side). It is common to use an IF of around 100 MHz. Because that is a convenient frequency, the LO commonly is 1880 – 100 = 1780 MHz or 1880 + 100 = 1980 MHz. (That choice means the image frequency is at 1880 + 1780 MHz and thus relatively easy to filter out.) By altering the output frequency of the local oscillator, different channels will be mixed down to the intermediate frequency, which is the mechanism for tuning to different carriers. For example, to tune to DECT RF carrier number 5 (1888.704 MHz), the LO is set to 1788.704 MHz; to tune to DECT RF carrier number 9 (1881.792 MHz), the LO is set to 1781.792 MHz.

The key parameter of the mixer is its linearity. Mixers, unfortunately, tend to be nonlinear devices; as well as producing outputs that are directly proportional to the input, they produce outputs proportional to the square, the cube, and so on, of the input signals. Cubing (raising to the power of 3) the input signals produces outputs at frequencies of the form $2f_1 - f_2$ and $2f_2 - f_1$, which lie in the operating band. Those outputs are called the intermodulation products. The magnitude of those unwanted products depends on the value of the input signal and the linearity of the mixer. As the magnitude of the input signal is increased, the magnitude of the unwanted component is cubed. In decibel terms, the intermodulation products increase 3 dB for every 1-dB increase in the wanted output. Because the unwanted signal output increases faster than the wanted output, there is a theoretical value of input signal at which the values of the wanted and unwanted signals are equal. That value is called the *intercept point*, and the linearity of the mixer is commonly quoted as the value of the third-order intercept point, or IP_3; the higher the intercept point, the better. That is why it is not prudent to increase the gain of the LNA stage excessively, since doing so will cause intermodulation in the mixer.

The DECT and PWT standards do not quote an IP_3 requirement directly. The DECT standard, for example, states that if the wanted signal is −80 dBm and there are two interfering signals each at −47 dBm, the receiver should continue to work successfully. Drawing the input/output graphs for the wanted and unwanted signals and simple trigonometry reveal that the value of intermodulation products can be estimated from the expression $V_{IMD} = 3V_{in} - 2IP_3$. Assuming that the required demodulator C/I is, say 10 dB, the interference signal must be less than or equal to −90 dBm (−80 dBm − 10 dB). Thus, the IP_3 must be −25 dBm or better.

The next device in the receive chain is a bandpass filter, which defines the channel and which is where the bulk of the channel selectivity is provided. The bandpass filter separates one received carrier from all the others and is at a fixed IF frequency. The key requirements of the bandpass filter are that it has sufficient bandwidth to cope with the wideband signal and does not distort the received signal. Surface acoustic wave (SAW) devices are a common choice. One of the most important things to consider in the design of the filtering arrangements for a TDMA/TDD system is that filters often react badly to the change of termination impedance that arises when circuits to which they are connected are turned on or off. That may give rise to a delay in reaching full operating performance when a receiver or transmitter is turned on, and the energy stored in a filter may reappear at unexpected times after a circuit is switched off.

The bandpass filter may be followed by various stages of IF amplification, which in a double-conversion superheterodyne receiver include a second mixing stage. The gain and noise figure of the amplifiers are selected in conjunction with the gain, noise figure, and IP_3 of preceding stages to give the required overall sensitivity. An important function of the amplification stage is to measure the received signal strength and to produce an RSSI proportional to the received signal strength. The RSSI signal is used to measure both the quietest channels to be used for communication and the strongest base station. The circuit must be linear over a range of 60 dB, as discussed in Section 4.4.1.

In a DECT receiver, the final amplifier in the chain may hard-limit the GMSK signal, since there is no information in the amplitude of that signal. The signal may be demodulated with a simple frequency discriminator. The digital data may be recovered using a clock recovery circuit and a bit slicing circuit, which compares the current signal value with the average signal value. The performance of the demodulator may be improved using a coherent receiver, which reproduces an exact replica of the carrier that was used to modulate the signal. However, such a demodulator adds to complexity and cost.

The data then are passed to the baseband digital and audio stages. Those stages typically are implemented in a burst mode controller IC responsible for all the data formatting, the ADPCM and PCM codecs, and the audio stages to drive the earpiece.

4.5.2 Typical Transmitter Structure

The transmitter in Figure 4.10 is the mirror image of the receiver. The speech is digitized and coded to the 32-kbps ADPCM format and packaged for transmission by a burst mode controller IC. The baseband data then are filtered before passing on to a modulator.

For the GMSK modulation used in DECT, the modulator may be as simple as using the filtered data signal to control the output of a voltage-controlled oscillator (VCO). Rather than using a separate VCO as a modulator, it is possible to modulate the VCO used in the synthesizer.

For the $\pi/4$ DQPSK modulation used in PWT, a quadrature modulator is used. Here the incoming data first are grouped into 2-bit symbols, then encoded as one of four phase changes between subsequent symbols, and used to feed in-phase and quadrature channels. That formally gives a complicated-looking expression for the magnitudes of the I and Q channels, which in reality means the I channel is proportional to the cosine of the phase angle of the current symbol and the Q channel is proportional to the sine. Each of the quadrature channels is then filtered (by a square root raised cosine filter) and mixed with in-phase and quadrature carrier components before being added in a summing amplifier.

After modulation, the signal may be mixed up to the operating band. Often, the same intermediate frequency used by the receiver IF is chosen so that the same synthesizer can be used to mix the signal to the RF frequency.

Finally, the power amplifier (PA) amplifies the signal up to its final output value. There is a considerable difference between the PAs needed for DECT and those needed for PWT. The amplitude of the modulated DECT signal is constant, which allows DECT equipment to use amplifiers operating in their most power efficient modes. Amplifiers are at their most power efficient when they are running close to their maximum output. However, as the PA output begins to saturate, its operation becomes nonlinear. That nonlinear characteristic will distort any amplitude component in the signal. Because a $\pi/4$ DQPSK-modulated signal used for PWT does contain amplitude variations, the amplifiers must be operated in their linear region. If the amplitude component is distorted, the spectrum occupied by the signal will be increased. One of the techniques for improving the linearity of an amplifier is through compensation. Compensation can be accomplished by predistorting the signal with a circuit that has the inverse response of the amplifier, so the final signal is undistorted. Alternatively, compensation can be accomplished using feed-forward techniques, in which the distortion is extracted, amplified, and then fed forward to be subtracted from the amplified signal. Because the amplified signal contains both the amplified wanted signal and the amplified distortion, subtracting a signal containing only amplified distortion leaves a pure undistorted signal.

A linearity requirement for the PA is implied by the limit on the transmitter intermodulation products. In single transceiver base stations, however, that limit is academic, since two or more signals are required to produce an intermodulation product.

In both cases, the PA must have pulse shaping to allow the transmitter power to be gently ramped on and ramped off. A sudden initiation of the transmitter burst will cause spectrum "splatter." A power-time template is defined for the transmitter to follow, to limit the amount of spurious signals that may be produced on other channels when the transmitter is switched on and off.

The PA output can be followed by a filter to remove unwanted emissions, although it is preferable not to use such a filter since its loss (approximately 3 dB) will require a correspondingly higher PA output power.

4.5.3 Receiver Sensitivity

Sensitivity is one of the most important receiver parameters. It is the receiver sensitivity in conjunction with the transmitter power that fundamentally determines the cell size that can be achieved.

Sensitivity is a function of thermal noise floor, the receiver noise figure, and the S/N required by the modulation scheme. The thermal noise floor is a physical constant that depends on the receiver bandwidth, the temperature in Kelvins, and Boltzmann's constant. The receiver noise figure is determined by the noise figure of the individual components and their respective gains and locations in the receiver chain. The S/N required by the modulation scheme depends somewhat on the type of demodulation circuit, with coherent demodulators giving better performance than noncoherent ones.

The minimum sensitivity for the receivers is set by the standards and the application profiles, to provide a uniform level of performance from equipment and to ensure minimum coverage areas for base stations and portables. That enables system operators to plan their base station deployments with the assurance that all handsets will obtain adequate performance throughout the planned coverage areas. The DECT and PWT standards require different values of sensitivity. The basic DECT requirement is for a sensitivity of –83 dBm. The DECT GAP interoperability standard requires –86 dBm, and the PWT requirement is for –90 dBm. The fact that sensitivity depends on bandwidth partly explains why the PWT sensitivity is lower than that of DECT. The narrower-band PWT system has a lower noise floor and thus may achieve better sensitivity. The difference between the DECT base standard and GAP requirements may be explained by the fact that the base standard gives merely a minimum figure, whereas the GAP requires a figure closer to realistic performance figures. The value of sensitivity quoted in the GAP is not necessarily state-of-art, as manufacturers' literature often claims around –90 dBm. However, the value required in the GAP in effect puts a cap on the maximum value of sensitivity. There is little benefit in developing products with improved sensitivity, since multicell networks must be planned to cope with all handsets on the market, including those meeting only the minimum requirement.

Within the standards, sensitivity is defined twice, once as the value of signal strength needed to achieve a minimum *acceptable* bit-error ratio (BER) and once as the value needed to achieve a *good* BER. The former is known simply as the sensitivity. The latter is the reference BER, which requires that the receiver can ultimately produce a good error ratio (better than 1 error in 10^5) when there is a good signal. That value sometimes is known as the irreducible BER, because increasing the signal strength beyond this level does not improve the BER.

For compliance with the standard, sensitivity is defined and measured under the unrealistic laboratory conditions of a nonfading channel. The performance of a product in real life depends strongly on its ability to cope with a rapidly changing radio channel. Sensitivity is measured most realistically when the receiver is operating in TDD mode, since there is often a big difference between practical receiver performance when it is operating simply as a receiver and when it is being switched rapidly between transmit and receive.

In many deployments, the practical range of DECT and PWT transmissions often are limited by interference from other users rather than by interference from thermal noise. The ability to tolerate interference from other users also has a significant effect on the coverage and the total system capacity. The tolerance to interference is expressed in terms of the C/I, which is a function of both the modulation scheme used and the type of demodulator. GMSK is robust to interference and the DECT standard requires operation at a C/I ratio of 10 dB (–73 – –83 dBm). For the higher order $\pi/4$ DQPSK modulation, the requirement is for operation at a C/I of 13 dB (–80 – –93 dBm).

4.5.4 Antenna Diversity

Transmission of signals through a radio channel is made more difficult by the fact that the radio channel is very unstable in space and over time. The signal strength at one point can be as much as 100 or 1,000 times (20 to 30 dB) weaker than that only a wavelength or so away. Such "fading" is caused by the transmitted signal reflecting from and bending (diffracting) around different objects and reaching the receiver by different paths. When multipath signals combine at the receiver, they may add to each other or cancel each other out. Also, each of the multipath signals takes a different length of time to travel from the transmitter to the receiver; if the difference in time is significant, it appears to the receiver as *time dispersion*, or echoes of the wanted signal. (The practical impact of such propagation effects is discussed in Section 8.3.)

The fading effects of multipath signals make it difficult to receive a radio signal reliably, and much of radio system design is devoted to overcoming that problem. Antenna space diversity is a highly effective method of combating the effects of a fading radio signal.

Antenna space diversity uses two antennas. Because the fading effect is highly dependent on position, if one antenna is in the depths of a fade, there is a good chance that the other is not. A typical scheme using antenna diversity to improve the received signal is illustrated in Figure 4.11. One of the key issues in antenna diversity is making sure that the signals received at both antennas are not correlated; therefore, the fading statistics at each antenna must be uncorrelated. That is achieved by spacing the antennas as far apart as possible, at least $\lambda/4$. For that reason, antenna diversity is not commonly implemented at the portable, since there usually is not enough room to allow the required physical separation between the two antennas. Within a portable, an alternative technique is to use polarization diversity. The two antennas are at the same location, but one is polarized to receive vertically polarized waves and the other

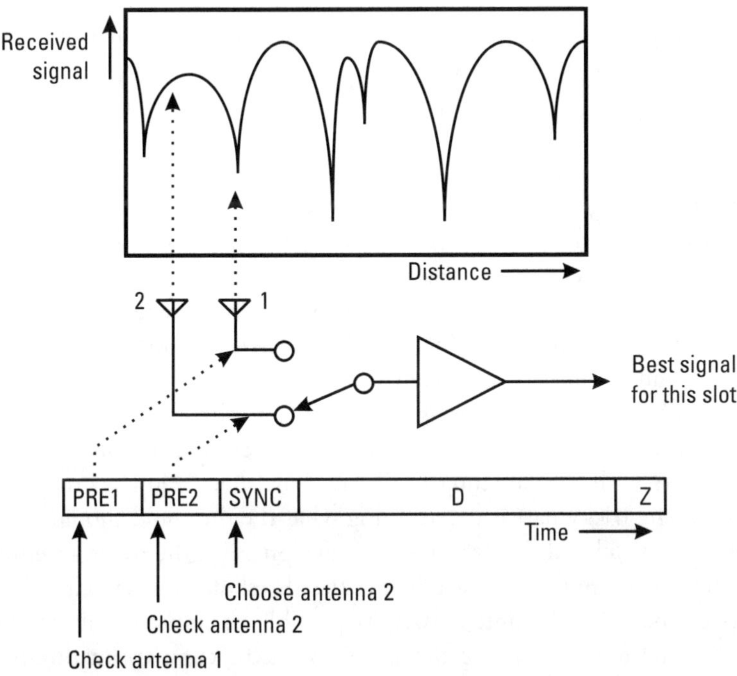

Figure 4.11 An implementation of antenna space diversity.

to receive horizontally polarized waves. This is quite effective for portables, because they normally are held at an angle of about 45° and the multipath reflections rotate the polarity of the received signal.

In theory, the fact that DECT and PWT use the same frequency to transmit and to receive enables some diversity gain to be achieved through the implementation of antenna diversity at the base station only. If the handset detects conditions under which it feels that the base station should switch the antenna on its behalf, it can request the base station to do that by setting a particular bit (the Q1 bit) transmitted in each slot. In practice, because the timeslots for receive and transmit are 5 ms apart, there is enough time for the multipath environment to change sufficiently that the uplink and downlink paths are somewhat different. Over 10 ms, the change may be rather substantial, and the effectiveness of the Q1 bit to signal the need for an antenna change is questionable.

In practice, however, there are alternative ways to take advantage of antenna diversity, including the following:

- Combining the signal from both antennas;

- Switching to one of the antennas and staying with that one until the signal becomes too weak;
- Choosing the better of the two antennas.

Combining the signal from both antennas can be effective but also complex. One of the best forms of diversity is maximal-ratio combining, in which the signals from the antennas are phase-shifted to bring them into phase with each other and then amplified in proportion to their signal strength, so that strongest signals are amplified most. Then all of the signals are added. This technique is complex and expensive for simple base stations.

One of the simplest methods of diversity and the most commonly implemented is switch-and-stay. In this method, the receiver selects one of the antennas and stays with that antenna until the signal falls below a predetermined threshold. The tricky part is determining what the threshold should be. If it is too low, the switch will ignore much better signals on the other antenna. If it is too high, if both antennas are below the threshold, the switch will switch repetitively between the antennas. It is possible to improve this method by using the long-term average of the signal to set the comparison threshold. In that case, the antenna is switched if the signal on the current antenna falls below the average. The amount of switching also can be reduced by taking the short-term average of the signal from the current antenna before deciding whether to switch. Setting up time constants and threshold levels can be tricky, especially for production volume equipment, and it is not unknown for a bad setup to make things worse. The fundamental problem with this method is that it does not test the alternatives before switching. Fortunately, the TDMA structure leads to a simple method of comparing the signal strengths on the two candidate antennas before making a decision.

The diversity performance is much improved if it is possible to compare the signal strengths of both antennas before deciding to switch. In a traditional receiver, that would require two independent receivers or at least two receiver front-ends, which could simultaneously monitor the strength on both antennas. However, because all transmissions are prefaced with a preamble and a synchronization field, it is possible to measure the signal on each antenna during this period and then to switch to the better of the two in time for the data field. A special mechanism exists to facilitate that. It provides the option of doubling the length of the preamble signal. The extended preamble gives the receiver more time to measure the signal on each antenna. It is not, strictly speaking, necessary to extend the preamble to measure the signal strength on both antennas. However, whether it can be done depends on the demodulator settling time and the time over which the channel must be measured to get a representative view of the signal strength. DECT provides a prolonged preamble of an extra 16 bits, and PWT provides an extended preamble of 24 bits. Note that switching between antennas should be synchronized with the bit clock so that antennas are not switched in the middle of a bit interval.

While diversity is effective in improving performance in noise-limited conditions, it is not as effective in interference-limited conditions. That is because signal strength conditions are reciprocal, but interference conditions are not. A base station may well experience interference not experienced by the handset since there may be an interferer close to the handset but not to the base station. Thus, channel selection is driven by the portable.

4.5.5 The Synthesizer and Blind Slots

The synthesizer is the device that determines to which RF carrier the receiver is tuned. It is a programmable oscillator that produces the LO frequency. The rate at which the synthesizer can change channels also determines whether the transceiver can use all 12 timeslots.

A typical synthesizer is a VCO whose output frequency is phase-locked to a stable crystal reference frequency. Because crystals do not operate directly in the operating frequency range, the synthesizer's output frequency is divided down by a set of dividers before being comparing with the reference frequency. To tune the synthesizer to different frequencies, the value of the divisor is changed, which produces an output from the comparator circuit, which causes the VCO to change to the new frequency. The output from the comparator is filtered through a lowpass filter before being applied to the VCO. That keeps the loop stable and limits the rate at which the output frequency can be changed from one carrier frequency to another.

The time taken by the synthesizer to settle on the new carrier frequency determines if the receiver can receive on different carrier frequencies on adjacent timeslots. A simple synthesizer settled on one carrier frequency may take several milliseconds to tune to and settle on a new carrier, whereas the guard space between timeslots is on the order of 50 μs. The exact time taken to settle depends on the bandwidth of the loop filter. To allow the receiver (or the transmitter) to use different radio carriers on adjacent timeslots, it is necessary to use either a fast switching synthesizer or a pair of synthesizers, one of which will be changing frequency while the other is stable. In that way, two slow synthesizers can be used to receive (or transmit) on adjacent timeslots.

The GAP requires that base stations can receive on at least 6 out of the 12 timeslots and that a portable can lock to any base station timeslot and transmit or receive on any frequency on any slot that is not immediately adjacent to a slot that the portable is using. That means a single-synthesizer system has to be capable of changing from one radio carrier to any other during, at most, the period of one timeslot (416.7 μs). If the synthesizer can meet that requirement but cannot change carriers within the guard period between slots, then it will have a number of *blind slots* (Figure 4.12). When it has settled on a carrier frequency during one timeslot (the filled squares in Figure 4.12), in immediately adjacent timeslots, it will be unable to use any other radio carriers (the plus signs in Figure 4.12).

Furthermore, if a base station has only one radio transceiver, then it cannot listen to two frequencies at once, which leads to further blind slots. That also can be seen in Figure 4.12 (the Xs), which shows that when the receiver is tuned to one carrier it cannot receive on other carriers. A portable cannot assume that if by measurement it finds a free slot on a particular carrier that it can use that to communicate with the base since the base already may be using that slot on a different carrier. To avoid that problem, the base station broadcasts a list of the nonblind slot pairs that it can use.

The existence of blind slots implies a reduction in capacity, since rather than the entire group of duplex 32-kbps physical channels being available (120 for DECT), first of all, only channels on idle timeslots are available. Then

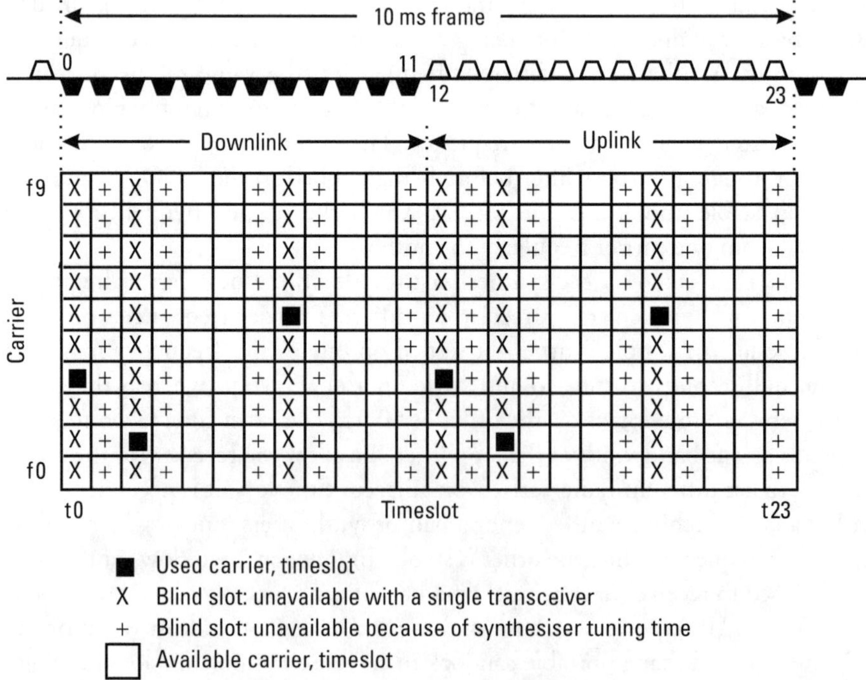

Figure 4.12 Channel availability with a single transceiver, with slow single synthesizer.

if the synthesizer is not fast enough, channels are available only as long as they are not adjacent to a used timeslot. If, for example, there are 10 carriers and a base station is using just 3 slot pairs (as in Figure 4.12), then rather than having a choice from the remaining 117 for new calls or handovers, a base station with a slow synthesizer may be able to choose from only 44 available duplex channels.

While that is not necessarily a problem in a single-cell system, in a multicell system many of the channels will be in use in neighboring cells and thus also unavailable. The number of usable channels for new traffic at a base station thus may be quite limited, and the use of a fast synthesizer usually is a good idea.

References

1. ETSI, *Digital Enhanced Cordless Telecommunications (DECT); Common Interface (CI); Part 2: Physical Layer (PHL),* ETS 300 175-2, Sophia Antipolis, France, September 1996.

2. TIA/EIA, *Personal Wireless Telecommunications Interoperability Standard (PWT); Part 2: Physical Layer (PHL),* TIA/EIA 662-2-1997.

3. TIA/EIA, *Personal Wireless Telecommunications—Enhanced Interoperability Standard (PWT-E), TIA/EIA 696-1996.*

4. Federal Communications Commission, *Unlicensed Personal Communications Service Devices,* CFR 47 part 15, subpart D.

5. Steer, D. G., "Coexistence and Access Etiquette in the United States Unlicensed PCS Band," IEEE Personal Communications, Vol. 1, No. 4, 1994, pp. 36–43.

6. Steer, D. G., "Synchronisation Interface for Low Power PCS Systems," *Proc. ICUPC '95,* Tokyo, 1995.

5

The Medium Access Control Layer

5.1 Introduction

The characteristics of the PHL have a profound impact on the way the protocol works as a whole, but it is the MAC layer that bears the brunt of its influence. So together, the MAC layer and the PHL form the real heart of the protocol.

The MAC layer provides a means for higher layer parts of the protocol to access the radio spectrum without knowing about its more intimate details. To do that, the MAC layer *does* know a great deal about the PHL, and it embodies a large number of rules about how to use the radio spectrum. The result is that the MAC layer creates a higher level model of the physical transport that is more closely aligned with the ultimate purpose: to provide cordless telecommunication services. The higher layers of the protocol, those above the MAC layer, complete the process started by the PHL and the MAC layer of adapting the radio medium to the task of carrying telecommunication applications.

The MAC layer is responsible for the operation of the radio spectrum just as we might monitor the spectrum with a spectrum analyzer or a radio receiver. It controls whether a cordless system is transmitting on certain timeslots and on certain radio carriers, whether it is receiving, or whether it does neither. It acts as a carrier of data. At the top of the MAC layer, it communicates through logical channels with the higher layers of the protocol stack and with the LLME, hiding the existance of data packets. It assembles data packets from those channels to be handed down to the PHL for transmission and performs the reverse translation for reception (Figure 5.1).

In most protocol stacks, there is a point where hardware implementation of the lower layer protocols gives way to software implementation of the higher

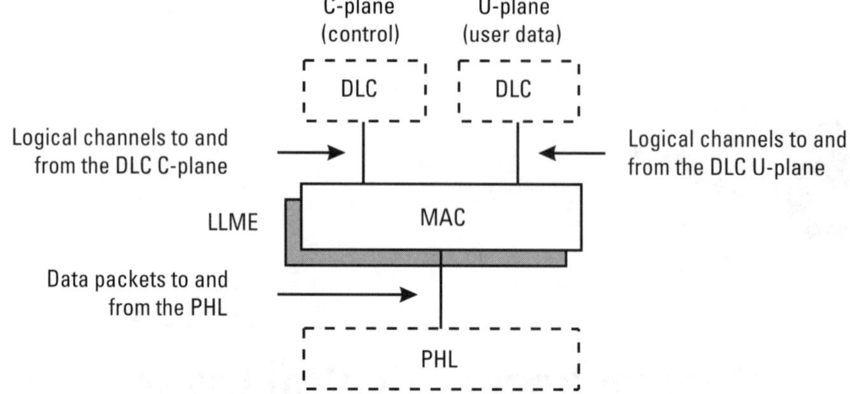

Figure 5.1 The MAC layer relative to the other protocol entities.

layers. That normally happens somewhere in the MAC layer. Most of the MAC layer's low-level data multiplexing (discussed in Section 5.4) is likely done best in hardware, in what usually is called a burst-mode controller chip. The high-level multiplexing, the various connection control entities, and the MAC layer's management entity normally are software tasks.

The DECT and PWT MAC layer standards [1,2] specify a large number of features, procedures, and data types, covering a wide range of possible applications. Thus, a MAC layer standard can look complex at first glance. One of the best ways to understand it is to look at it from the point of view of a specific application. Once the correct versions of the all data structures, procedures, and services are found and the others eliminated, you are left with a more concise and consistent set of MAC layer features that can be more easily understood.

Section 5.2 covers the basic concepts of the MAC layer, including brief descriptions of all its important features. The rest of the chapter looks at the MAC layer as a voice service carrier in a system that conforms to the GAP [3]. The GAP is an application-specific profile that specifies which features of the MAC layer (and other layers) are required to provide voice telephony. In PWT, the equivalent access profile for speech services is the CPAP [4]. The CPAP has slightly different requirements compared to the GAP. Where those differences are significant, they will be highlighted.

If you take a copy of the MAC layer standard and highlight those parts listed in the GAP or CPAP, then highlight in a different color the parts of the MAC layer that are not, you will find that many pages of the MAC layer are needed only for data transport, packet-mode operation, or other more advanced systems. (Chapter 8 deals with more advanced applications according to other DECT service profiles.)

5.2 Basic Concepts

If we could say just three things that the DECT MAC layer does, they would be these:

- The MAC layer creates bearers from the data packets it sends to and receives from the PHL. It then builds the bearers into connections that the higher layers of the DECT protocol can use to make telephone calls and obtain other telecommunications services.
- The MAC layer also creates connectionless and broadcast services from its bearers, including a beacon full of information, broadcast from every radio cell in every fixed part (FP), to allow any handsets in the vicinity to find out all they need to know to gain access to the system.
- Finally, the MAC layer assigns the contents of the PHL's data packets to a number of logical channels. Those channels have specified function and meaning, and the MAC layer exchanges them with higher layers of the protocol stack, from DLC upward, and with the LLME.

To do those tasks, the MAC layer introduces a number of key concepts and functions that are the fundamental ideas provided by DECT to elaborate and specify its services. You need to understand those ideas and the MAC layer's organization to understand its operation in detail.

5.2.1 Bearers

Using the PHL's highest level abstract model of the DECT/PWT system, you can send and receive data packets of several different sizes on any of a number of physical channels. The packets are called P00, P08j, P32, and P80 (see Section 4.3.2). The first level of the MAC layer's functionality allows the packets to be used to create bearers. A continuous simplex bearer is created by sending or receiving one PHL data packet in every 10-ms TDMA frame. A continuous duplex bearer is created from two simplex bearers in opposite directions 5 ms apart. The larger the packet used, the wider the data channel created. The PHL is, in fact, completely idle unless the MAC layer is operating and is ordering it to transmit and receive data packets. It is, therefore, the MAC layer that creates bearers by regularly sending packets of data to the PHL or by regularly requesting the PHL to receive them.

The DECT bearers, therefore, provide either simplex or duplex variable bandwidth data channels (or bit pipes) between the fixed and portable DECT parts. The normal duplex bearer for a GAP system, to take just one example, transmits and receives one P32 packet per TDMA frame. That provides a 32-kbps duplex path for the user's data channel plus a separate 6.4-kbps duplex data channel for control and signaling. That particular bearer allows ADPCM-encoded two-way voice telephony to be carried.

5.2.2 Connections

Having built the concept of bearers on top of the PHL's packets, the MAC layer then builds the concept of a connection made up from one or more of the bearers. A connection is what the DLC layer requests from the MAC layer to create a link, which the NWK layer then uses to create a telecommunications service such as a telephone call. Figure 5.2 illustrates the entire hierarchy from packet to bearer, then connection, link, and finally call.

In the MAC layer, connections first have to be established, then they provide data transfer, sometimes with flow control and/or error protection; they are released when no longer needed. Connections always use that three-phase process.

In the establishment phase, just after the DLC layer asks for a connection to be made, the MAC layer, in communication with the MAC layer at the

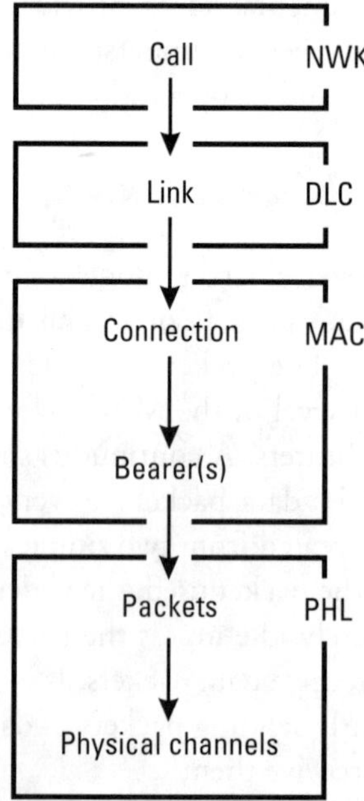

Figure 5.2 The hierarchy from call to physical channels.

other end, allocates one or more bearers to each connection it establishes. In the simplest case, a single duplex bearer may be allocated to the connection, to provide normal voice telephony. In more complex cases, to support full 144-kbps ISDN 2B+D, for example, several duplex bearers may be allocated to the connection to provide the necessary bandwidth.

The MAC layer, therefore, builds its connection-oriented services from one or more of its various bearers. The bearers are in turn built from the PHL's data packets. A versatile set of different connection-oriented services of various data rates, both symmetrical and asymmetrical, can be obtained in that way.

5.2.3 Broadcast and Connectionless Services

As well as building connection-oriented services from its bearers, the MAC layer's simplex bearers can be used to build what are called connectionless services. A connectionless service has no setup or release phase. It has only a data transfer phase. Connectionless bearers, therefore, may suddenly appear, broadcast their information, then disappear when no longer needed. Unlike connection-oriented services, connectionless services are received only on a prearranged basis. Recipients must be continuously expecting data broadcasts to appear and disappear without being explicitly prepared at the right time through a connection setup process.

Connectionless services are used either for broadcasting information to multiple recipients or for sending data packets to destinations that are expecting to receive them. That allows the protocol to create broadcast beacons and fast packet-based data services, to give just two examples.

5.2.4 Channel Allocation and Handover

The MAC layer allocates the PHL's physical channels to bearers on a dynamic basis. It has procedures in which the portable parts (PPs) and the FPs periodically create and maintain maps of the signal strength they receive on each of the PHL's physical channels. The MAC layer also establishes the rules by which PHL channels are selected and used. In most cases, if a bearer is needed for some connection, the physical channels with the lowest received signal strengths are selected. Such procedures allow for the maximum efficiency of spectrum use without the user or manager of a cordless system having to plan spectrum usage cell by cell.

The physical channels that carry the data for a connection may from time to time suffer interference. Also, if the portable is mobile, the received signal strength from a fixed station may become too weak to support a connection. In such cases, the MAC layer provides what is called bearer handover. In that process, the bearers underlying a connection are changed without the higher layers of the protocol, the DLC and NWK layers, being aware of the change. If the requirement to change bearers is detected early enough, the MAC layer can provide seamless handover by establishing a new set of bearers for a connection before the error rate on the old bearers has increased to a level where the user would notice. Having established the new bearers, the MAC layer transfers the connection, without a break in service, from the old bearers to the new ones.

The MAC layer also supports connection handover, which is a process by which the DLC changes to a completely new MAC layer connection to support a higher layer service. Connection handover is provided by the DLC layer for situations in which the cordless system becomes too large to be controlled by a single MAC layer and therefore too large to operate with just bearer handover. It also deals with any failure of the MAC layer to maintain a successful connection.

5.2.5 The Beacon

A fundamental service that DECT and PWT provide is a broadcast beacon. It is difficult to overestimate the influence the beacon has on the operation and performance of the system. The beacon normally is broadcast from the FP. Actually, the beacon can be transmitted independently by a portable to allow direct portable-to-portable communications, but that is not required by the GAP.

The beacon is a service created by the MAC layer. The MAC layer ensures that the beacon is broadcast in each cell of a fixed system (Figure 5.3). The beacon broadcast is carried alone if required, on its own special bearer. However, it also is appended to every continuous bearer carrying a user's traffic and so may be broadcast several times in each cell, whenever there is a lot of user traffic.

One of the functions of the MAC layer's beacon is to broadcast the fixed system's identities to all the portables in range. Portables are programmed in advance, through a subscription process, with the identities of systems that will provide them service.

The beacon broadcasts a number of important details about the fixed system's technical capabilities and the services it can provide. Handsets are expected to be able to use that information to adjust their behavior to work properly with the fixed system. For example, they may not invoke certain optional services unless the FP broadcasts that it supports those services.

The beacon also broadcasts paging information, the main purpose of which is to alert a portable that an incoming call is available. All basic MAC layer connections (including GAP voice connections) are ultimately set up by the portable, although some advanced connections may be set up from the FP.

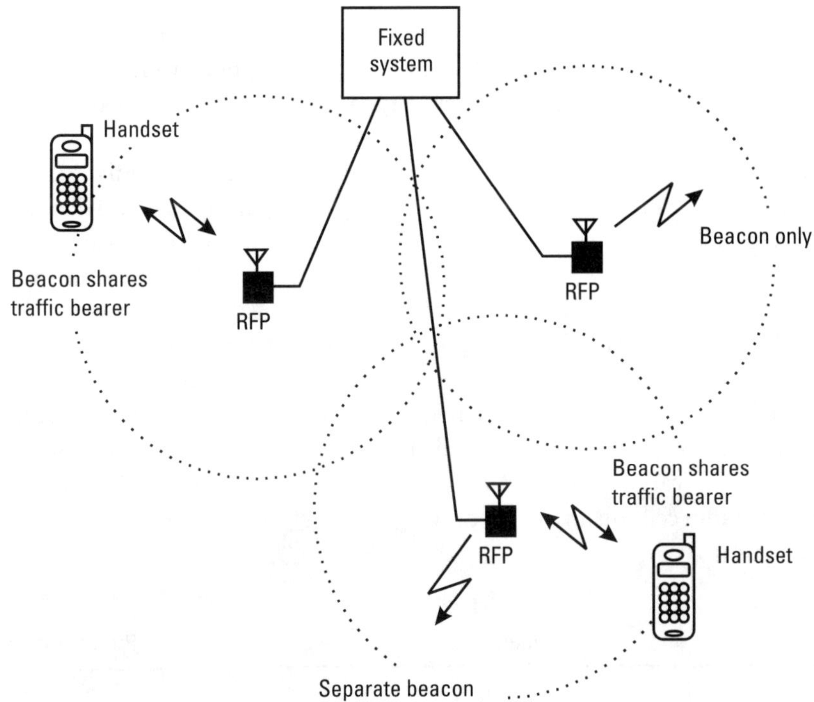

Figure 5.3 The beacon.

For incoming calls originating at the FP, a basic connection setup is achieved by paging the destination portable using the paging channel on a beacon and asking it to set up a connection. In that case, a connection is set up just to alert the user. However the speech channel is connected only if the user answers the call.

When a beacon is broadcast in conjunction with a normal connection, it shares the bearer's signaling channel. The beacon coexists with the internal messages between the peer MAC layers at the two ends of the connection and with the signaling data between higher layers of the protocol. The higher layer signaling includes signaling associated with the user's connection and signaling for other purposes, such as handset location tracking. A process called the T-MUX coordinates the multiplexing of all of that information.

5.2.6 Cells and Clusters

The MAC layer creates the concept of clusters of radio cells. The rest of the standard allows higher levels of system organization involving the operation of many clusters, whenever a system is too big for one MAC layer implementation to support (Figure 5.4). Since a cluster of cells may support from 1 to 255 cells, a single MAC layer can support a very large DECT system if needed and if it is convenient.

Some of the MAC layer's functions are associated with individual radio cells (formally called radio fixed parts, or RFPs), which operate independently in each cell. Those functions are called cell site functions (CSFs). Bearer control and receiver control are examples of functions that are associated with a specific cell site and that operate independently.

Some of the MAC layer's functions, however, are shared among many cells, which is what, in effect, defines a cluster. Those functions are called cluster control functions (CCFs). Generally, the connection-oriented, broadcast, and connectionless services are handled at a cluster level. That is because a specific communication of one of those types may use the services of a number of different radio cells during its lifetime.

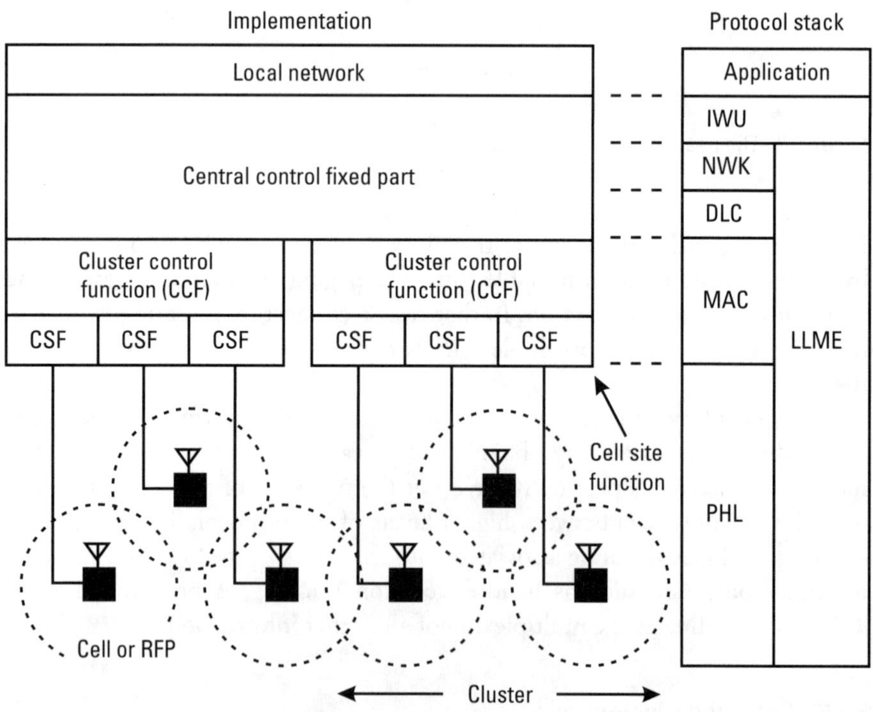

Figure 5.4 Clusters of cells and higher level organization.

5.3 Basic Organization and Operation

Before we describe in more detail the way in which services are provided by the MAC layer, we have to establish a formal framework for the MAC layer's organization and its operation. That will be done first by introducing the reference model of the MAC layer provided in the DECT and PWT standards and then by describing the formal details of the MAC layer's bearers and connections.

5.3.1 Reference Model

As we have already seen, the MAC layer is organized into two sets of functions. The first set comprises functions that are needed on a cell-by-cell basis and that are concerned with low-level matters, such as the operation of individual bearers. The second set comprises those functions that control the operation of a cluster of cells and that are concerned with higher level matters, such as controlling connections. Hence, we can see two groups of functions within the MAC layer: the lower group, made up of the CSFs, and the higher group, made up of the CCFs (Figure 5.5). Communication with the world outside the MAC layer takes place through SAPs, and the chunks of data that are exchanged are SDUs.

You can see from Figure 5.5 that the job of the CSFs is to communicate directly with the PHL in the various cells of a cordless system. That is done through what is called the D-SAP, of which there is one per cell. The D-SAP carries data for or from the D-field of a PHL packet (see Section 4.3.4 and Figure 4.7). At the higher level, the CCFs' communication with the DLC layer takes place through three more SAPs, the MA-SAP, the MB-SAP, and the MC-SAP, which carry different logical channels for different purposes. The CCFs also communicate with the LLME through the ME-SAP.

The SAPs are there for reasons of convenience and formal identification. They may or may not be used explicitly in real implementations. Remember that the MAC layer's internal structure as described here does not have to reflect the exact physical or logical organization of any specific system. However, there normally is a correspondence between the MAC layer's internal reference model and the "usual" organization of a multicell DECT system.

We come now to the full MAC layer reference model, shown in Figure 5.6, which illustrates the three specific CCFs and the four specific CSFs.

All the MAC layer functions can be thought of as independent processes. Whenever a connection or a service is required by a higher layer protocol, one of the CCFs is started up and ordered to do a specific job and then report the result. It then may create and communicate with CSFs,

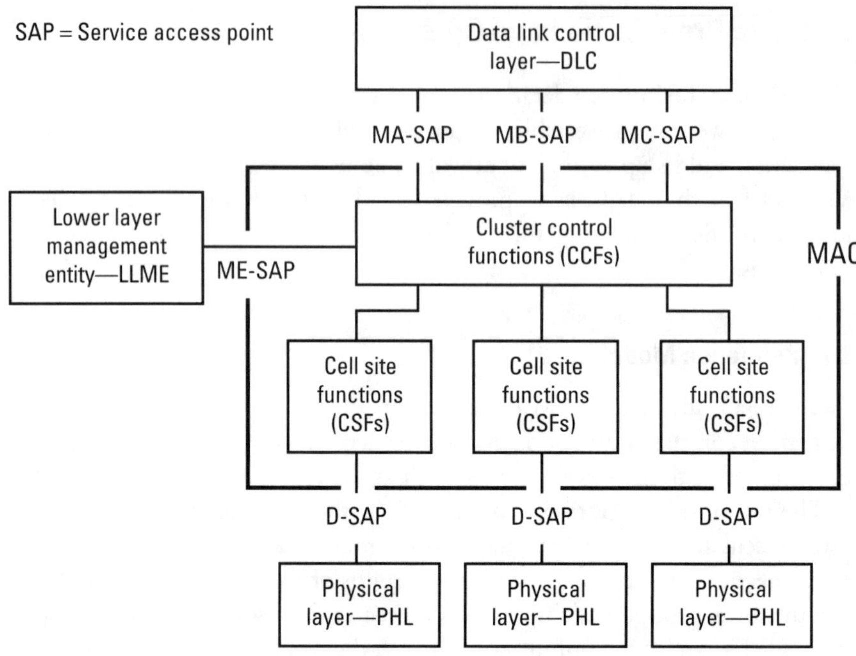

Figure 5.5 The basic MAC layer organization (one cluster). *Source:* ETSI [1].

ordering them to take on subtasks and report their results. When each job is complete, for example, a call is finished, the functions that were created to do the job are terminated.

In addition to the MAC layer's functions, Figure 5.6 shows some of the logical data channels used by those functions (which are covered fully in Section 5.4.2).

5.3.2 The Cell Site Functions

Generally, as is clear from their individual names, the CSFs concern themselves with the management of bearers. Even the idle receiver control (IRC), in spite of its name, is concerned with bearers; it organizes a cell's unused receivers to scan for incoming bearer setup requests.

A cell comprises one or more radio transceivers, each offering access to all the physical channels of the DECT or PWT system. With the use of TDMA and TDD, the cell, even with just one radio, usually is able to support more than one connection to local PPs. That means there may be more than one instance of a CSF in each cell. The CSFs are listed next:

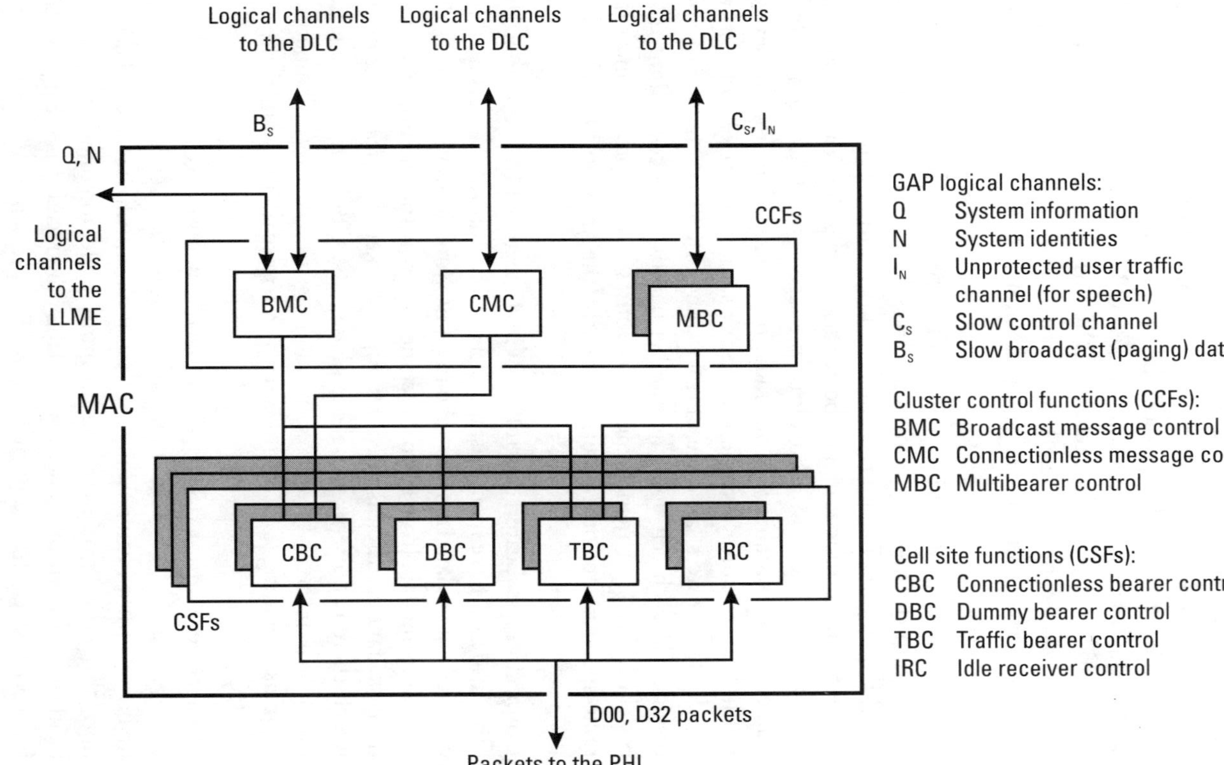

Figure 5.6 The MAC layer reference model. *After:* ETSI [1].

- Dummy bearer control (DBC). The DBC creates and controls a continuous bearer for an independent beacon transmission, that is, a beacon transmission not associated with a bearer carrying user traffic. This special bearer is called the dummy bearer. The DBC is controlled by and gets all of this information from a broadcast message control function at the cluster level. Up to two beacons are allowed to be associated with a single cell and, hence, up to two DBCs. When certain types of user traffic are present in a cell, the DBC may be ordered to release its dummy bearer, because the beacon data from the broadcast message control function already are being transmitted with the user traffic on other bearers.
- Traffic bearer control (TBC). The TBC controls the operation of a connection-oriented bearer used for voice or other continuous duplex data. The user traffic and control signaling it carries come from a multibearer control function. The bearer also carries the beacon transmission, taking data from a broadcast message control function. In one cell, even if it contains only one radio transceiver, there may be several traffic bearers, since a single transceiver sometimes can handle up to 12 duplex bearers at once. Hence the TBC may have multiple instances in each cell site. The number of TBCs in a cell is limited by certain MAC-layer management procedures, for the protection of the radio spectrum for other users. Currently, the limits in DECT and in PWT are different.
- Connectionless bearer control (CBC). Like the TBC function, the CBC also may have many instances per cell. It controls bearers used for connectionless data transfer. Like the TBC function, it also carries the beacon using data from a broadcast message control function. Connectionless data transfer is used for rapid packet data transmission in which there is no setup or release phase. The CBC simply broadcasts the data it receives from a connectionless message control function. Systems providing GAP-compliant voice service do not require the use of CBCs.
- Idle receiver control (IRC). There may be several IRCs per cell, one per transceiver. The IRC function controls all receivers not already involved with a bearer. Whenever a cell has an idle receiver, the IRC normally sets that receiver into scanning mode, looking for incoming bearer-setup attempts. The first available idle receiver is set up to perform what is called the primary scan. The primary scan is a specified sequence for scanning the physical channels of the DECT or PWT system. If there are one or two more idle receivers in a cell, they are set up

to perform the secondary and tertiary scans in the same sequence as the primary scan but lagging a specified amount of time behind it.

5.3.3 The Cluster Control Functions

There can be from one to 255 cells in a MAC cluster, but only one group of CCFs is associated with a single cluster of cells. As with the CSFs, there may be several instances of some of the CCF functions within the cluster. Complete systems are not limited to one cluster, but performance and cost benefits sometimes can be achieved by implementing a system as a single cluster.

- Broadcast message control (BMC). There is only one BMC per cluster, and it operates the services used on a beacon. The BMC's services are attached to all existing continuous bearers, with each bearer carrying similar messages. If a bearer does not exist within any individual cell, because there is no traffic being carried, then the BMC always creates a dummy bearer to broadcast the BMC's services on a downlink beacon (fixed to portable) without any traffic attached. The BMC obtains system information (the Q-channel) and system identities (the N-channel) from the LLME and takes paging information (the B_S-channel) from the DLC layer. The detailed rules for transmitting a beacon differ between DECT and PWT.
- Connectionless message control (CMC). The CMC, of which there is at most one per cluster, controls all connectionless point-to-point or point-to-multipoint services on connectionless bearers in both downlink and uplink directions. An example of the services controlled by a CMC is a packet data service. A system conforming to the GAP standard does not require provision of a CMC.
- Multibearer control (MBC). There may be several MBCs per cluster, each controlling the traffic bearer(s) for a connection-oriented service, such as a speech connection. Systems conforming to GAP use the MBC for their voice connections. In GAP systems, the MBC obtains slow control signaling (the C_S-channel) and user traffic (the I_N-channel) from the DLC layer.

5.3.4 The MAC Layer's Bearers

Bearers are basic bit pipes, created by the CSFs driving the PHL and causing it to transmit and/or receive packets of data. Bearers carry data across the air interface. There are only four defined types: the simplex bearer, the duplex bearer, the double simplex bearer and the double duplex bearer (Figure 5.7). They are created, managed, and released at each cell site by the CSFs.

A group of one or more of the bearers makes up a connection or carries the data for a connectionless service. Combining the bearers allows a versatile set of connections to be created.

- A simplex bearer comprises one physical channel at the PHL, and transmissions occur in one direction only. There are two types. The short simplex bearer uses only the PHL's P00 packet (see

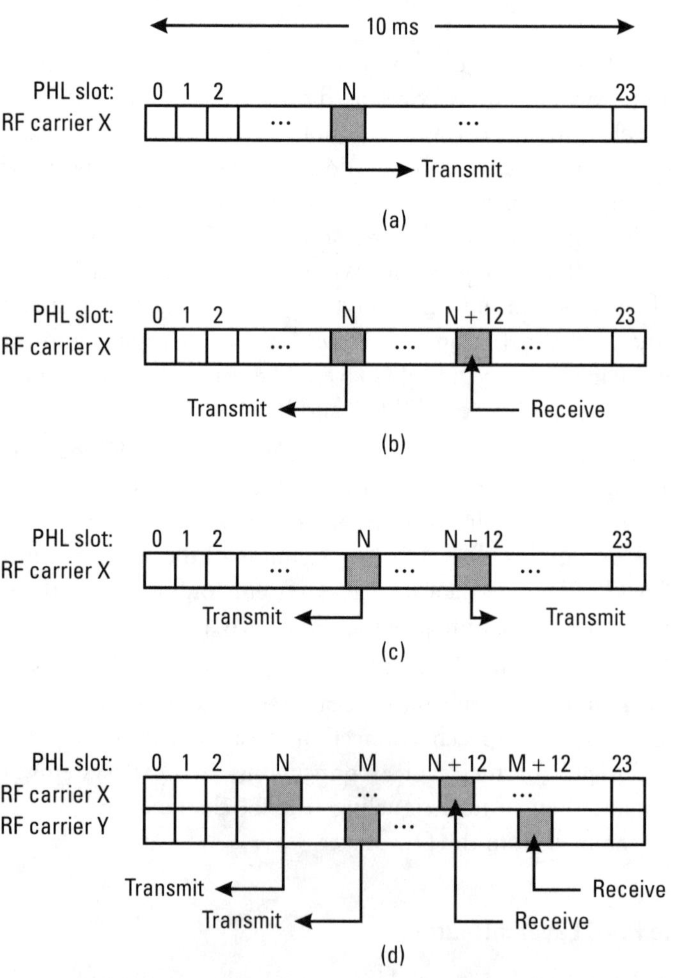

Figure 5.7 The MAC layer's bearers: (a) simplex; (b) duplex; (c) double simplex; (d) double duplex.

Section 4.3.2) and is controlled by a DBC function to make a stand-alone beacon transmission. The long simplex bearer uses other, longer PHL packets; it is created and controlled by a CBC function for connectionless services, such as data packet transmission. The short simplex bearer under the control of a DBC function is required by the GAP. The long simplex bearer under the control of a connectionless bearer control function is not required by the GAP.

- A duplex bearer is a pair of simplex bearers operating in opposite directions, using two physical channels at the PHL. The pairs of channels are always on the same RF carrier frequency and are always spaced half a TDMA frame (5 ms) apart. A duplex bearer can be created by a CBC function or by a TBC function. The duplex bearer using P32 packets, controlled by a TBC function, is used for GAP speech connections.
- A double simplex bearer is a pair of long simplex bearers operating on a pair of physical channels at the PHL, in the same direction. Like duplex bearers, they always operate on the same RF carrier and are always spaced half a TDMA frame apart. The double simplex bearer is always part of an MBC service. One double simplex bearer is controlled by one TBC function but is not required in GAP-compliant systems.
- A double duplex bearer is a pair of duplex bearers under the control of the same MAC connection. Therefore, there are always two TBC functions involved under the control of one MBC function. High-data-rate duplex traffic uses these bearers, but the GAP does not require them.

The four types of bearers are often said to be in one of three operational states, according to their use: dummy bearer, traffic bearer, or connectionless bearer.

- Dummy bearer. This label is attached to a short simplex bearer being used in a continuous transmission (i.e., one packet per frame) in the downlink direction, that is, from the fixed termination (FT) to the portable termination (PT). This is the stand-alone beacon transmission, created by a BMC function when there is no other traffic to carry. From the FT, the dummy bearer broadcasts paging messages and related information from the DLC layer, system information and system identities from the LLME, and internal MAC control messages.
- Traffic bearer. The bearer is being used for continuous point-to-point transmission under the control of an MBC function. A traffic bearer is

a duplex bearer (GAP) or a double simplex bearer (non-GAP). Remember that broadcast information from a BMC function is also multiplexed onto a traffic bearer, and a traffic bearer may take the place of a dummy bearer for broadcasting the beacon, whenever there is traffic to be carried.

- Connectionless bearer. The bearer is controlled by the CMC function and carries packets of data. As with the traffic bearer, it also must contain broadcast information from a BMC function. The connectionless bearer may operate in either the downlink or the uplink direction, and it may be either simplex or duplex. This is not a bearer category required for GAP speech service.

The MAC layer transmits a bearer by sending to the PHL a PL_TX-req primitive, indicating the type of bearer required and containing the data for one packet. If the desired bearers are continuous, the MAC layer has to send a new primitive in every frame. The MAC layer receives a bearer by sending the PHL a PL_RX-req indicating the type of bearer to receive, and the PHL replies with a PL_RX-cfm containing the data, an indication of whether synchronization was achieved and optionally containing other information about the quality of the communication channel.

5.4 Multiplexing and Messages

The MAC layer performs a rather complex multiplexing procedure to allocate the data bits in PHL packets to the various logical information channels. As part of the multiplexing scheme, the MAC layer also handles the encryption of certain information. The MAC layer multiplexer also carries certain logical channels that exist only within the MAC layer, to carry certain MAC layer-to-MAC layer information and the messages needed to set up, maintain, and release the MAC layer's bearers.

The protocol stack distinguishes two fundamental types of data. The first type is the user data that the system has to transport to justify its existence. An example in a GAP-compliant voice system is the coded voice itself, for which the GAP allocates a 32-kbps channel in each direction on a duplex bearer, called user-plane (U-plane) data. The second fundamental data type is the control signaling that the protocol needs to control its own operation, called control-plane (C-plane) data. These names conveniently classify the various logical data channels, which are described next in more detail.

The general structure of the MAC layer multiplexer is shown in Figure 5.8. Most of, if not all, the lower levels of multiplexing, including the CRC generation and checking, are in the hardware of a real implementation, a burst-mode controller chip.

5.4.1 Packet Structure

Working from the PHL upward in Figure 5.8, we see that the PHL calls the packets of data it exchanges with the MAC layer "the D-field." At its lowest level, the MAC layer defines what it calls the "D-MAP," which divides the D-field into the A-field (which is always 64 bits) and the B-field (which may have four different lengths, from 0 to 804 bits). Hence, the data packets are of different lengths, depending on application and service.

The four different D-fields (called D00, D08j, D32, and D80 by the MAC layer) correspond to the PHL's packets, P00, P08j, P32, and P80. The packets occupy an A-field only, a half slot, a full slot, and a double slot, respectively, at the PHL. Note that the D08j field is extensible and so can use up the guard space between slots when that is necessary, but the current standards do not define any permitted extension over the 148-bit nominal size. The idea is, in the future, to find a low-rate speech codec to double the capacity of the system, which might need to occupy a little more than the 8 kbps that the P08j packet would supply with $j = 0$.

For GAP-compliant voice systems, only two specific D-fields (i.e., two packet types) are relevant. The field called D32 is used in the P32 packet for carrying 32-kbps digital voice in the B-field plus control and beacon data in the A-field. The D00 field (in the P00 packet) carries control data only in the A-field and is used to form the independent beacon on a dummy bearer.

5.4.2 Logical Information Channels

At its highest levels, the MAC layer deals in logical channels, not data packets. Logical channels are used as a shorthand method of classifying and describing the information flowing through the protocol stack. There are actually two types of logical channels: logical channels that exist outside the MAC layer and the MAC layer's own internal logical channels.

The protocol defines several external logical data channels that just flow through the MAC layer. In that case, the MAC layer takes no notice of the data it carries. The MAC layer's actions with respect to those data channels are limited to retransmitting any data that are in a protected channel whenever a transmission error occurs. Any data in an unprotected channel that get corrupted are lost.

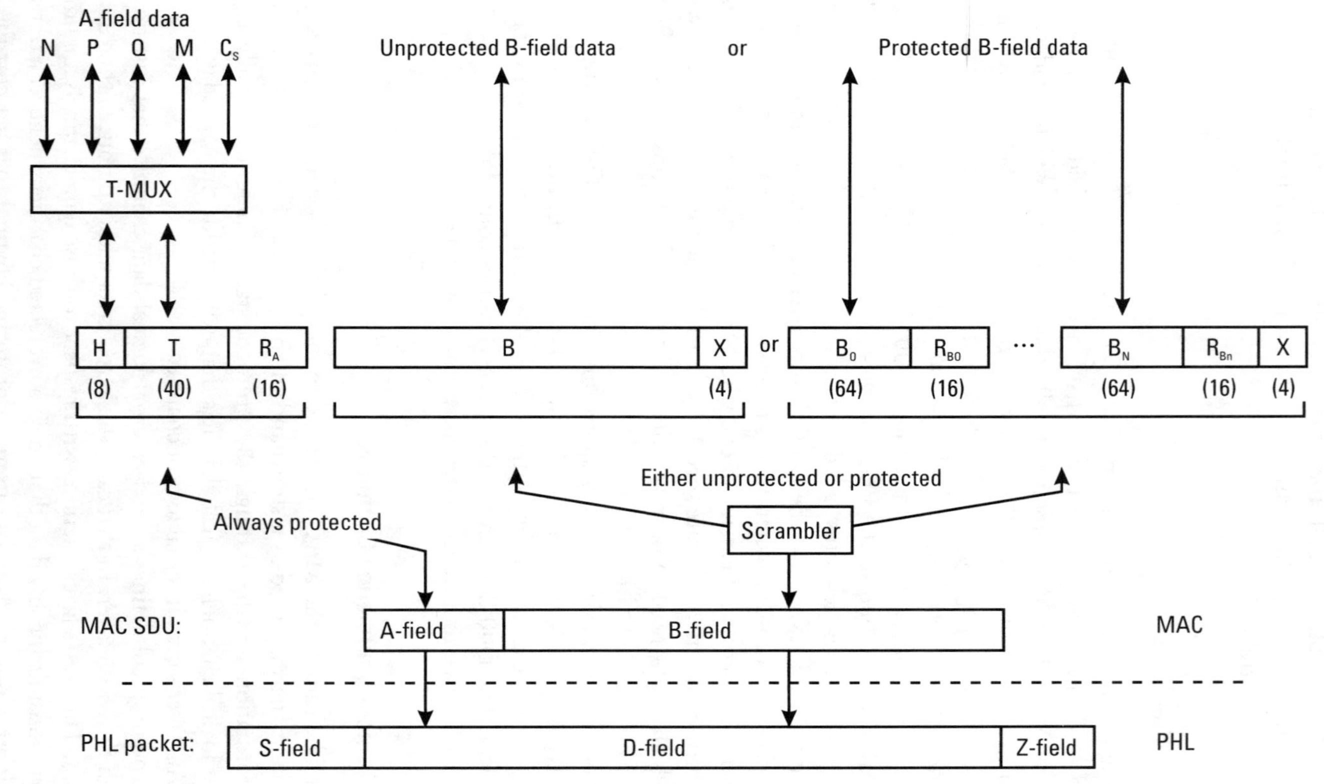

Figure 5.8 The general MAC-layer multiplexing scheme.

In the case of the MAC layer's own internal logical channels, called the M-, N-, P-, and Q-channels, the other parts of the DECT protocol stack are not aware of them, although the lower layer, the PHL, carries the data in its packets. The internal MAC channels are protected channels in that errors may be detected, but retransmission may not be requested by the recipient. In the normal course of events, the MAC layer ensures that transmissions on the internal channels are repeated as required until no longer needed.

5.4.2.1 The I-Channel

The I-channel is the U-plane data, for example, digital speech. In the transmit direction, it comes into the MAC layer from the DLC layer and is passed down to be inserted into the B-field in PHL data packets.

Speech data normally use the I_N-channel, which is the name for the unprotected I-channel. In this context, unprotected means that no high-performance CRC protection is added before transmission, so most of the available bits inside a PHL packet can be used for the data. That maximizes the information flow, but if data are corrupted in transmission, they are lost. Actually, a small amount of protection is provided in unprotected channels (with the 4-bit X-field). That fairly weak CRC check allows most errors to be detected at the receiving end, so the system can take action such as muting the audio link if the data are digital speech. I_N is the channel used by the GAP for its speech service.

The MAC layer has two ways of collecting I_N-channel data from the DLC layer and inserting them into a packet (Figure 5.9). The first is to collect the data arriving up to the end of the last half of the 10-ms TDMA frame, insert that into a PHL packet, and then at the other end of the link unpack it and send it out to the user at the start of the next TDMA half-frame. This is called the I_N_normal_delay service, as shown in Figure 5.9(a), and presents the fewest system timing problems. However, it does add extra delay to the data channel, having about a 15-ms overall transmission delay (1.5 frames).

The alternative to I_N_normal_delay service is I_N_minimum_delay service, as shown in Figure 5.9(b), in which the data collected from the MAC layer are one frame's worth of data right up to the moment of collection. At the receiving end, the data are passed to the DLC layer and on to the user immediately. Although that minimizes the channel's delay to about 10 ms, several interlayer timing and synchronization issues must be resolved. Nevertheless, delay is an important issue in speech services, so GAP systems use the I_N_minimum_delay service.

There is also a protected I-channel, called I_P, that is inserted into the B-field of PHL packets with error-protection information. That reduces the available user bandwidth but allows errors to be detected at the receiving end, which

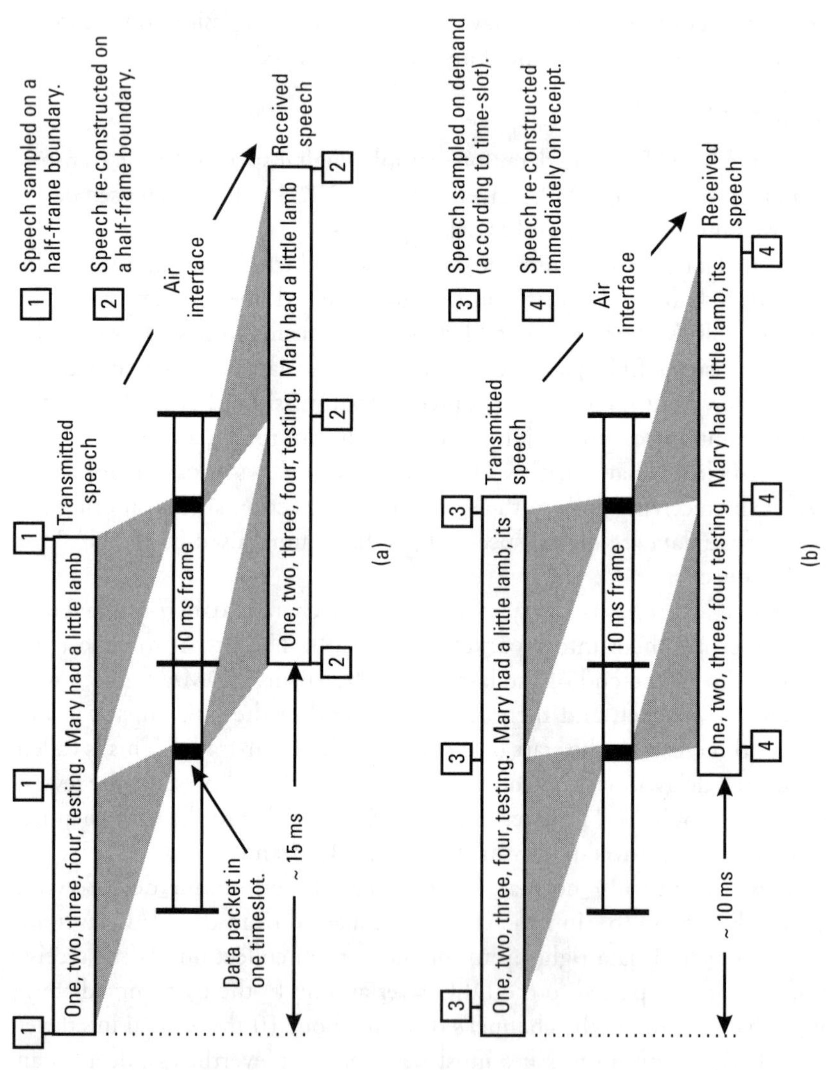

Figure 5.9 Speech transmission in (a) I_N_normal_delay service and (b) I_N_minimum_delay service.

then can request the I_P data to be retransmitted. That means that I_P data are always error free, but they may not flow in a predictable real-time fashion. For that reason, it normally is used for nonspeech services that do not object to such means of transmission.

5.4.2.2 The C-Channel

The C-channel is control information that originates at higher layers of the DECT protocol stack, in the DLC or NWK layer, or higher still in an application using the cordless protocol. An example might be a message to indicate that the user has pressed a key on a portable terminal, or an attach message that originates within the protocol. The MAC layer always protects that information and allows the receiving end to request retransmission using the MAC-layer ARQ procedure whenever an error occurs.

The C-channel has two varieties, the fast C-channel (C_F) and the slow C-channel (C_S), depending on how the MAC layer transmits it. The C_F-channel is fast because it uses the PHL's B-field and is allowed to interrupt I-channel (user) traffic. Thus, it is not suitable for use with real-time speech connections and is not allowed in speech systems conforming to the GAP standard. In GAP systems, the C_S-channel is used instead. The C_S-channel is allocated space in the 64-bit A-field of a PHL packet, but there is not so much bandwidth there, so it is rather slower than the C_F-channel. However, there is enough bandwidth available using the C_S-channel for all normal signaling for a telephone call.

Note that although the C-channel is part of the C-plane, it is not the same, in spite of the similarity. C-plane information includes the C-channel, but the C-plane also includes much more.

5.4.2.3 The G_F-Channel

The G_F-channel is a fast (i.e., B-field) simplex data channel that is used for the control of U-plane entities at the higher protocol layers. For example, it can be used to carry acknowledgments in asymmetric connections. It is not required in a GAP speech system.

5.4.2.4 The SI_N-Channel and the SI_P-Channel

The SI_N-channel is unprotected connectionless U-plane data that use the B-field. It is the connectionless equivalent of the I_N-channel and shares many of its facilities, including a limited error-detection coding (just the 4-bit X-field without retransmission facilities). Like all connectionless services, it is sent out to any receiver that is listening at the time. The SI_P-channel is protected connectionless user data. Unlike the SI_N-channel, it is inserted into the B-field with substantial error-protection information (several 16-bit CRC fields). Either channel could be used for a connectionless speech transmission, but neither is required for the GAP speech service.

5.4.2.5 The C_L-Channel

The C_L-channel is used for connectionless C-plane data. Just as for the C-channel and for the same reasons, it exists in slow and fast versions: CL_S and CL_F. The GAP specifies connection-oriented speech services only, so it does not use the C_L-channel.

5.4.2.6 The B_S-Channel

The B_S-channel is the slow broadcast channel; there is no fast equivalent. This channel is used only in the downlink direction, from FT to PT, and typically carries call setup requests and called paging messages, asking relevant PTs to respond. Other uses are allowed. The B_S-channel is always protected by the MAC layer in such a way that transmission errors can be detected. However, the MAC layer does not allow the receiving end to request retransmission. Normally, B_S-channel information is rebroadcast anyway, so a receiver usually will be able to pick it up the next time around if there is an error.

5.4.2.7 The M-Channel

The M-channel contains MAC layer–to–MAC layer signaling messages that are created by and used within the MAC layer. This channel, therefore, is used to communicate between the MAC layers of two cordless parts, for the purposes of bearer setup, for example. The M-channel is required by GAP (and for all applications). It is error protected with automatic retransmission. The MAC layer's own procedures, which use the M-channel for signaling, deal with any loss of M-channel information.

5.4.2.8 The N-Channel

In all applications, system identities are broadcast on the beacon by FTs and PTs in the N-channel. In general, all RFPs broadcast N-channel information on all traffic, connectionless, and dummy bearers they have, to form part of the DECT beacon. PTs broadcast N-channel information on their traffic bearers. The information originates at the LLME. When a PT sets up a bearer, it echoes back the N-channel information it receives from the FT, in effect performing a MAC handshake.

5.4.2.9 The P-Channel

The P-channel is the paging channel, which is broadcast by FTs on the beacon. It contains B_S-channel data from the DLC layer, which is used to broadcast paging messages for portables. It also contains other MAC information related to bearers and bearer setup.

5.4.2.10 The Q-Channel

The Q-channel is the system information channel and is broadcast on the beacon. The information originates from the LLME. The Q-channel assigns frame numbers, modulo 16, so creating a 16-frame multiframe cycle, which is used as part of the A-field multiplexing scheme, the T-MUX (see Section 5.5.3). It carries a fairly large amount of information about the operation and capabilities of the cordless system.

5.4.3 The A-Field

The A-field (see Figure 5.8) is present in all PHL packets. Used only to carry C-plane data, it is divided into the header, tail, and redundancy (CRC) fields. The redundancy bits here, as elsewhere, are calculated from the rest of the field's data in a specified way for the purposes of error detection or correction.

The header of the A-field is used to designate what information channel is currently carried by the tail; to designate what is in the B-field; and finally to transport various quality indications (Table 5.1). The quality indicators include data on CRC failures in both the A-field and the B-field and other indications about the air interface, such as the need for antenna diversity change and the onset of interference from other asynchronous DECT systems whose frame structure is drifting through.

Table 5.1
Structure of the A-Field Header

Field	Bits	Purpose
TA	3	Set via the T-MUX (see Section 5.5.3) to indicate the type of data in the A-field tail: C-, N-, Q-, M-, P-channel or proprietary. GAP supports all values except proprietary.
Q1	1	For bearer quality control in duplex connection-oriented bearers. Sometimes called BCK.
BA	3	To indicate the type of data in the B-field. Only two types of B-field data are supported in GAP: BA = 000 for I_N (U-plane) data and BA = 111 for no B-field.
Q2	1	For bearer quality control and C-channel flow control in duplex connection-oriented bearers.

The information in the A-field tail is formatted as MAC messages (see Section 5.4.5). There are two sets of those messages, one intended to be transmitted in the A-field tail, which is required for GAP support (Table 5.2), and another set intended to be sent in the B-field tail, which is not required by GAP systems.

5.4.4 The B-Field

The B-field can contain two types of data: signaling data or user data. The exact type of data in the B-field is indicated by the BA-field within the A-field header. This multiplexing arrangement between signaling and user data is called the E/U-MUX.

Whenever the B-field is allowed to carry signaling (E-type) data, control signaling is always given precedence over user (U-type) data, and user data may be temporarily interrupted or delayed. The use of E-type signaling data in the B-field is specified by what is called the C-MUX. The C-MUX can be used for fast control signaling between the cordless parts. For speech services, however, interrupting the speech channel for control data is not acceptable, so control data in the B-field are not allowed in GAP. Because GAP does not use signaling in the B-field, the C-MUX is not required and is not covered here. However, Table 5.3 does list the data that can be carried by the C-MUX.

The only B-field size supported in GAP is the 324-bit B-field inside the D32 D-field. This B-field contains 320 bits of user data plus four bits of CRC generated from the user data. It is transmitted in a P32 packet at the PHL every 10 ms, providing the 32-kbps channel for U-type (typically speech) data.

Table 5.2

Information Multiplexed Into the A-Field

Information	Source	Purpose
C_S-channel	DLC	Slow higher-layer control signaling, C_S, or slow connectionless control signaling, CL_S (note that support for C_S is required under GAP, but not CL_S).
M-channel	MAC layer	Internal MAC layer-to-MAC layer control signaling
N-channel	LLME	System identities
P-channel	DLC (FT only)	Portable paging messages (which contain B_S-channel from the DLC and other MAC-layer information)
Q-channel	LLME (FT only)	System information

Table 5.3
Signaling Information in the B-Field

Information	Source	Purpose
C_F-channel	DLC	Fast higher-layer fast control signaling, C_F, or fast connectionless control signaling, CL_F
G_F-channel	DLC	Fast higher-layer user control signaling, G_F
M-channel	MAC layer	Internal MAC layer-to-MAC layer control signaling

5.4.5 MAC-Layer Messages

Structures called MAC-layer messages are used to carry all the C-plane data to and from the PHL. The messages themselves are created and managed by the MAC layer; however, they carry data from, or for, the MAC layer and the LLME, as well as data that originate in or are destined for the DLC and higher layers. MAC-layer messages can be sent in either the A-field or the B-field of a PHL packet. GAP systems support only A-field messages, so (for now) we can safely ignore the entire section of the MAC standard devoted to B-field MAC-layer messages.

The MAC layer's beacon is created by broadcasting certain MAC-layer messages in the PHL's A-field, containing identities (N-channel), paging (P-channel), and system information (Q-channel).

MAC-layer messages also carry MAC layer–to–MAC layer (M-channel) communications that contain bearer setup, maintenance, and release messages, among others. They also carry higher-layer control data (C-channel), with non-GAP fast signaling (C_F) carried in B-field messages and slow signaling (C_S) carried in the A-field. Both M- and C-channel messages are addressed to a specific recipient rather than being broadcast. When messages are sent in the A-field, they are sent in the A-field tail, and the type of message is indicated in the TA-field of the A-field header (see Section 5.4.1 and Table 5.1). When the header indicates that a particular logical channel is contained within the tail field (e.g., the N-channel), then the tail usually is labeled according to that channel (the N-channel tail in this example).

The GAP standard requires support for all the MAC-layer message types but only some of the individual MAC layer messages. Marking up a copy of the MAC layer standard by reference to those clauses of the GAP (or CPAP) standard that list the required messages is the quickest way to determine which must be supported and any special considerations required.

5.4.5.1 N-Channel Tail

The primary function of the N-channel tail in an FT is to broadcast an identity message (Table 5.4). This message contains a radio fixed part identity (RFPI), which is a 40-bit identity. It is constructed mostly from a primary access rights identifier (PARI), as well as the radio fixed part number (RPN), which is an identity of up to 8 bits that uniquely identifies a specific cell within a fixed system. No other messages are specified. The PARI and all the RPNs are identities assigned by a system manufacturer and/or installer. (Details of how identities are assigned are covered in Section 8.4.)

The primary function of the N-channel tail in a PT is during a connection to rebroadcast the RFPI that a portable has received from an FT, thus performing a MAC identities handshake.

5.4.5.2 Q-Channel Tail

The Q-channel tail broadcasts mostly the Q-channel system information listed in Table 5.5. However, it does sneak in some N-channel (identities) information in the form of the secondary access rights identity (SARI) list. More information about how SARIs are assigned is given in Section 8.4.

5.4.5.3 P-Channel Tail

Paging channel messages are defined of several lengths, depending on the number of bits required to identify the recipient (Table 5.6). GAP requires only short and zero-length paging messages.

The zero-length page message does not page any particular handset, but it is used to broadcast information about how portables should respond to paging messages. The short page message carries the 20-bit temporary portable user identity (TPUI) that is assigned to a handset by the higher layers of the protocol. The assignment is made when the PT has locked to a system and has sent a NWK-layer attach message to indicate its presence. More details about that and the other mobility management processes are covered in Section 7.6.

Table 5.4
N-Channel MAC-Layer Messages

Message	Required by GAP/CPAP?
Identities	Yes

Table 5.5
Q-Channel MAC-Layer Messages

Message	Required by GAP/CPAP?
Static system information	Yes
Extended RF carriers	Yes
FP capabilities	Yes
Extended FP capabilities	No
SARI list	Yes
Multiframe number	Yes, but only if encryption is supported
Escape	No

Table 5.6
P-Channel MAC Layer Messages

Message	Required by GAP/CPAP?
Full/long page message	No
Short page message	Yes
Zero-length page message	Yes

5.4.5.4 M-Channel Tail

The M-channel tail is primarily for internal MAC layer–to–MAC layer control messages used for such purposes as connection setup and maintenance (Table 5.7).

The GAP standard requires support for just the basic connection control message set, which includes these messages:

- access_request;
- bearer_handover_request;
- connection_handover_request;
- bearer_confirm;
- wait;
- release;
- unconfirmed_access_request.

Table 5.7
M-Channel MAC Layer Messages

Message	Required by GAP/CPAP?
Basic connection control	Yes
Advanced connection control	No
MAC-layer test messages	No
Quality control	Yes
Broadcast and connectionless services	No
Encryption control	Yes
First B-field setup message	No
Escape	No
Tertiary access rights identity (TARI) message	No
REP connection control	No

The function of the basic connection control messages is covered in Section 5.6, including the use of the fixed MAC identity (FMID) and the portable MAC identity (PMID), which appear in most of them.

5.5 The Beacon

The beacon broadcasts a range of information from the FP to all PPs in range, allowing listening portables to lock their behavior to that of the FP and gain access to the other services from the beacon and from the system. It provides three basic services to PPs:

- Access rights information (N-channel and Q-channel);
- System information (Q-channel);
- Paging information (P-channel).

It already has been stated that it is difficult to overestimate the influence of the beacon on the operation and performance of a DECT or PWT system. The beacon is the means by which portables find out about their environment, so they know where in a system they are or to which system they are nearest. It informs them if the system will accept their calls. It is the means by which they synchronize timing with the FP and are able to turn off their electronics for long periods, so saving battery power without the danger of missing incoming calls. It is the means to deliver notification of incoming calls, and finally it informs the portables of the capabilities of the fixed system, so they can adjust their behavior accordingly.

5.5.1 Creation of a Beacon

All RFPs within a system must broadcast a beacon in one of several ways. When a system is turned on, its first function is to establish beacons in all its cells. Also, whenever a new cell comes on-line, the first thing it does is to establish its beacon. Under certain circumstances, two beacons are allowed in a cell, provided they support the same services.

The beacon always uses the A-field of a PHL packet to transmit its information (see Section 5.4.3), and the information is multiplexed onto the A-field using the T-MUX (described in Section 5.5.3). It is broadcast in one (or more) of three ways:

- On a dummy bearer. If the system has no continuous bearers in place, it has to create at least one dummy bearer containing an A-field only (using the PHL's P00 packet). Broadcasting a beacon on a dummy bearer is required by the GAP standard.
- On a connectionless bearer. If the system has one or more continuous connectionless downlink (fixed to portable) bearers, the A-fields of those bearers are used for the beacon. The GAP service, however, does not require support for connectionless services.
- On a traffic bearer. If the system has one or more traffic bearers in place (such as those used for GAP speech), the A-fields of those bearers must be used for the beacon as well as carrying the connection's signaling. The B-fields of the bearers are used for user traffic.

Whenever there is a lot of traffic between a cell and surrounding portables, there may be a choice of many beacons for an otherwise idle portable to scan (Figure 5.10).

All the beacons in a cell attempt to carry the same broadcast data for all portables. However, in traffic bearers and connectionless bearers, the A-field in the PHL's packets is shared by M-channel and C-channel messages destined for a specific portable. Those channels get priority over broadcast data if there is any conflict (see Section 5.5.3), so the various beacon transmissions are not always identical. A dummy bearer broadcasting the beacon never has to make

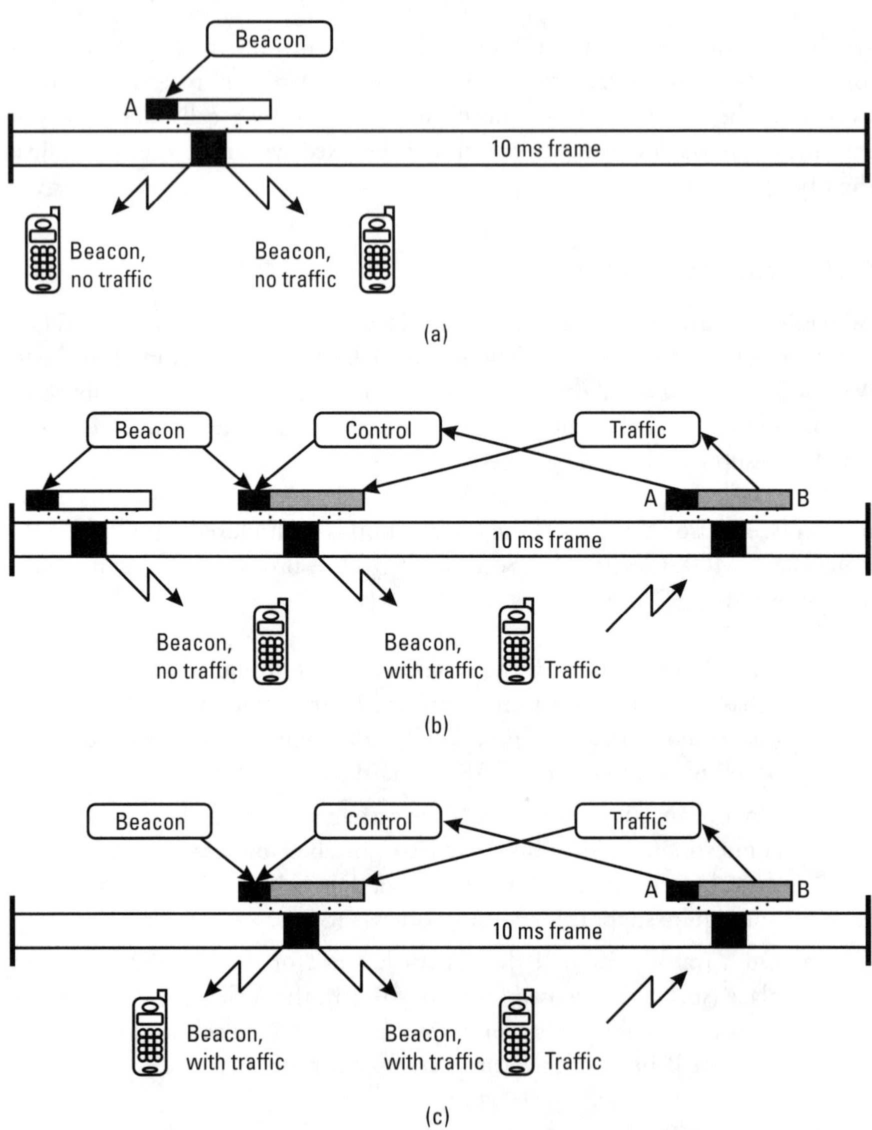

Figure 5.10 (a) Beacon only; (b) separate beacon and traffic; (c) combined beacon and traffic.

room for other data channels and so provides the quickest means for a portable to acquire all the broadcast data. However, that normally is not a significant difference. Note also that C-channel messages sharing A-field capacity with beacon broadcasts will contain private information, such as dialed numbers. If privacy of information is essential, the C-channel may be encrypted along with the user's speech or data on the traffic bearer.

Formally, under the MAC layer's reference model, it is the continuous broadcast procedure, controlled by the BMC and using the services of either the CBC, the TBC, or the DBC that creates the beacon.

5.5.2 Services Broadcast on the Beacon

Although it may share bandwidth with other information channels, the basic beacon broadcasts three types of information: access rights, system information, and paging.

It is the responsibility of the FP's LLME to make available to the MAC layer all the access rights information and all system information. It does that whenever required, by issuing via the ME-SAP a MAC_ME_INFO-ind primitive containing system identities (PARI, SARI, and RPN; see Section 8.4), FP capabilities, and the current multiframe number. The FP's MAC layer may respond with a MAC_ME_INFO-res primitive to confirm reception, and then it stores the information for use in all future broadcast messages. The procedure used for broadcasting is that of the T-MUX multiframe multiplexing scheme described in Section 5.5.3, using messages described in Section 5.4.5.

A system may have many identities. Its main identity, the PARI, is broadcast in the N-channel most frequently. The SARIs are broadcast less often in the Q-channel. The PP decodes from the various ARIs whether it has access rights to the FP in question (see Section 8.4).

System information is required by PPs to lock their behavior and adjust their timing to align with the fixed system. System information identifies the specific frame numbers within a multiframe and allows the portable to find out such things as how often paging messages are sent, what sequence is used by the FP to scan for bearer setup requests, and whether the FP can provide certain optional features or services.

The paging broadcast procedure is covered in Section 5.5.6. It is the T-MUX, however, that makes space available in the MAC-layer messages for two things: paging identities, which are there to alert portables that they have an incoming call, and special MAC-layer information, which tells portables how to respond if they are paged.

5.5.3 The T-MUX

The T-MUX lies at the heart of signaling in GAP equipment, both for the beacon and for call-related messages. Its primary purpose is to multiplex a series of information from the MAC layer and from other sources onto the A-field of a dummy bearer or a traffic bearer (see Table 5.2). This is the process that creates the beacon. Often, there is too much information to broadcast, so the T-MUX has a priority scheme that allows variable amounts of information of any given type to be broadcast, depending on demand. It specifies a minimum rate of information flow for each of the low-priority channels, so that less critical information is never completely squeezed out.

As part of the T-MUX, the MAC layer imposes a 16-frame multiframe period of 160 ms on top of the PHL's 10-ms frame. Internally, the frames are numbered 0 to 15. The data put into the head and tail subfields within the A-field depend on the multiframe number, the data available, and the channel's priority.

There are two T-MUX algorithms, one at the FP and one at the PP. That may seem strange, because the PP normally does not broadcast a beacon. Actually, the PP does not normally broadcast a dummy bearer, but whenever it has an active traffic bearer, it does transmit information in the A-field. Of course, it does not normally broadcast system information (Q-channel) or paging (P-channel).

The priority scheme in the T-MUX is specified for each frame. It may be specified, for example, like this: "M_T, C_T, N_T." This specific example means that M-, C-, and N-channel information may be broadcast in the A-field tail in this frame and that M-channel information has the highest priority, followed by C-channel information, followed by N-channel information with the lowest priority. In this case, N-channel information will be broadcast only if no C-channel or M-channel information needs to be sent.

5.5.3.1 T-MUX at the Fixed Part

The T-MUX at the RFP allocates a few special frames to make sure that certain information has a minimum frequency of broadcast (Table 5.8). Frame 8 is always allocated to the Q-channel (system information), and frame 14 is always allocated to the N-channel (identities). The position of the Q-channel uniquely in frame 8 is the key that allows a portable to obtain multiframe synchronization. The other even frames broadcast identities by default, but if there is paging to be done, it is given priority. In the odd-numbered frames, the broadcast of identities also is the default, but higher layer control information and MAC-layer messages are given priority.

There is one exception to the rules. When responding to a "bearer request" message or during bearer release, the FT may insert an M_T tail in an even-numbered frame.

5.5.3.2 T-MUX at the Portable Part

The portable uses the T-MUX algorithm on all its traffic bearers in connection-oriented services. It is simple (Table 5.9): odd frames always

Table 5.8
The T-MUX at an RFP

Frame(s)	Priority Scheme
0, 2, 4, 6, 10, 12	P_T, N_T
8	Q_T
14	N_T
1, 3, 5, 7, 9, 11, 13, 15	M_T, C_T, N_T

Table 5.9
The T-MUX at a PP

Frame(s)	Priority Scheme
0, 2, 4, 6, 8, 10, 12, 14	M_T, C_T, N_T
1, 3, 5, 7, 9, 11, 13, 15	N_T

broadcast N-channel (identities) information; in even frames, the priorities (from lowest to highest) are identities, then higher layer control signaling, and finally MAC-layer messages.

As with the FT, there is an exception. The transmission of a "bearer request" or a "bearer release" from a PT may use an M_T tail and can be placed in any frame.

5.5.3.3 Implementing the T-MUX

The most future-proof way of implementing the T-MUX is to be capable of receiving and processing any information in any frame, while transmitting according to the rules. The only thing that should be assumed at the portable is that the Q-channel can appear only in frame 8. However, because of the exceptions, that does not mean you can assume the Q-channel will appear in every frame 8. There are many places in the standards where, like here, only the transmission rules are specified. Implementers should be careful not to apply transmission rules to their receivers, or they may fall foul of potential future upgrades in the standard. The receiver should make no unecessary assumptions about the multiplexing algorithm, while the transmitter should enforce the rules.

5.5.4 Portable Part and Fixed Part States

The MAC layer introduces various states into the operation of the portable and fixed systems.

Perhaps the most important state machine is the one in the portable, in which the handset monitors the radio interface for the fixed system's beacons and when it recognizes an FP where it has been granted access rights, it downloads all relevant system information and uses this to move from an unlocked state to a locked state.

In the locked states, the portable knows from the system information broadcasts when to listen for paging messages indicating incoming calls, and it also knows when the FP's receiver is listening for its outgoing calls. That allows the portable to turn itself off for a large fraction of the time, so preserving battery capacity, while still being able to promptly receive incoming calls. It also allows the portable to correctly predict when to start setting up a new call.

5.5.4.1 Portable Part States

The PP in a DECT or PWT system aims to know all about the fixed systems in its environment, their system parameters, their capabilities, and their identities. Those pieces of information are broadcast on the fixed system's beacons. In that way, a portable can synchronize its behavior on a recognized fixed system. In principle, the portable could lock to more than one system, but if the systems' frames are not fully synchronized, keeping independent timings for the two systems' multiframes is difficult. Normally, therefore, a portable locks to only one beacon and then receives service from only that system. To formally describe this behavior, the PP MAC layer exists in four major states (Figure 5.11):

- *Idle unlocked* is the state in which the PP is not synchronized to any RFP and makes no attempt to gain lock with a suitable RFP. It is effectively the state of the PP when it is switched off or in some other standby mode in which it does not expect to make or receive connections.
- *Active unlocked* is the state into which the PP moves on power-on or activation from a standby mode. It is also the mode in which the PP remains if it cannot recognize an RFP that it can access, or if it loses lock to a suitable RFP. In active unlocked mode, it is the aim of the PP to find an RFP it recognizes and to receive enough information to lock to that RFP and enter the idle locked state.
- A PP enters the *idle locked state* when it recognizes an RFP and has extracted enough information about the RFP to know how to operate

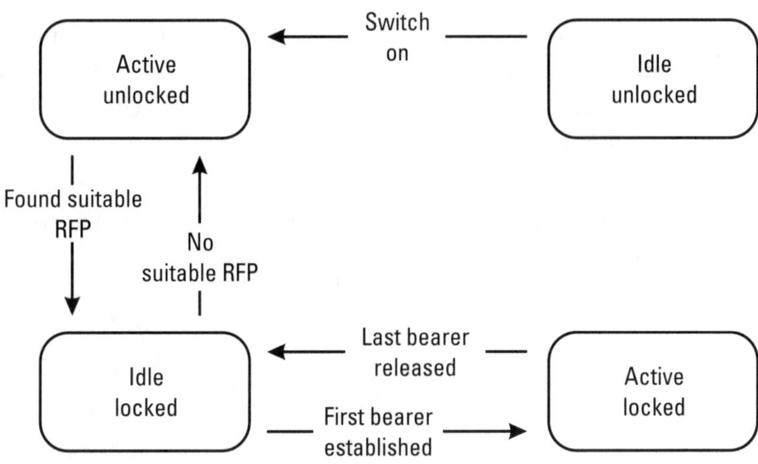

Figure 5.11 PP state diagram. *After:* ETSI [1].

with it and to know that it is allowed to make or receive calls. In this state, the portable is locked to the fixed system. There are four substates of idle locked mode:

Scanning mode is the mode in which the PP has determined the FP's receiver scanning sequence and has locked its own scanning sequence to the FP. That means the PP knows when to transmit information so the FP will receive it.

Normal idle locked mode is the mode in which the PP has locked its receiver to the FP's multiframe and wakes up just once per multiframe (once per 160 ms) to look for paging information (i.e., call-setup attempts). This mode of operation is required by GAP once a portable has become locked. The portable is able to shut down for most of the 160-ms multiframe and looks for incoming calls for only a short time within the multiframe. By doing that, the portable can have up to a 160-ms delay in recognizing an incoming call, but it saves a great deal of power, thus prolonging its battery life.

High-duty cycle idle locked mode is the mode in which the PP has locked its receiver to the multiframe of the FP and wakes up six times every multiframe to look for paging information. In this mode, the call-setup time is minimized at the expense of increasing the power consumption of the handset. However, GAP handsets do not have to support this mode of operation. The FP has to cooperate with this mode by not waiting for the normal position within the multiframe period to start its broadcast of a paging message with notification of an incoming call. Typically, this mode is used by data terminals.

Low-duty cycle idle locked mode is similar to normal duty cycle version, but the PP does not wake up every multiframe. Instead, it wakes up every four multiframes. In this mode, the PP power consumption can be reduced greatly, but the call-setup time is lengthened proportionately. This mode is not required by GAP.

- *Active locked* is the state in which the PP is synchronized to an RFP and has one or more bearers established. The PP enters this state when the very first bearer of the first connection is established and leaves the state when the last bearer of the last connection is released.

5.5.4.2 Fixed Part States

The RFP MAC layer exists in one of four major states (Figure 5.12):

- *Inactive* is the state in which the RFP is either not powered up or neither receiving nor transmitting.

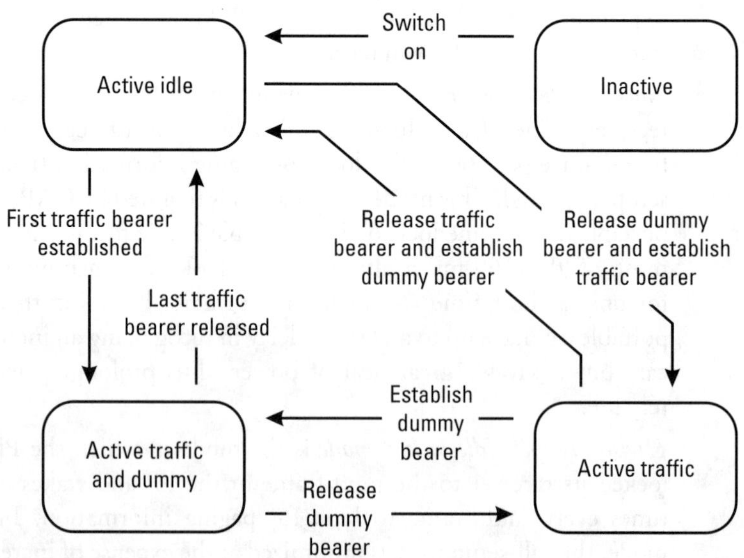

Figure 5.12 FP state diagram. *After:* ETSI [1].

- *Active idle* is the state in which the RFP has established at least one beacon but does not have any traffic. It uses either a dummy bearer or at least one connectionless downlink bearer. In this state, the RFP's receiver is scanning in the sequence advertised on the beacon channel.
- *Active traffic* is the state in which the RFP has at least one traffic bearer on which it also is broadcasting the beacon, but it does not have a separate beacon transmission.
- In *active traffic and idle*, a combination of two states, the RFP has a separate beacon on a dummy bearer and also maintains a beacon on all its traffic bearers and all its downlink connectionless bearers.

5.5.5 Portable Part/Fixed Part Locking Procedure

When a PP starts operating, for example, after it is first turned on, it moves from the idle unlocked state to the active unlocked state. In the latter state, it tries occasionally to enter the idle locked state by gaining synchronization with a recognized FT. To do that, the PT occasionally scans all physical channels for the beacons of an FT. The timing of attempts is not specified and is left up to the PT's LLME, considering matters such as power consumption and service availability.

Portions of the scan can be accomplished by the MAC layer sending a PL_ME_SYNC-req primitive to the PHL asking it to gain slot timing synchronization with the next FT. Then the PT MAC layer captures the beacon's transmissions with PL_RX primitives and decodes the N- and Q-channel information. The Q-channel information enables the PT to obtain multiframe synchronization and determine the FT's primary receiver scanning sequence.

The PARI and the SARI from the beacon are passed on to the LLME for checking, and the LLME may then inform the MAC layer that an acceptable system has or has not been found. If the FT is recognized, the PT's MAC layer then continues the scan, looking for the specific RFP within the system that has the highest signal strength. The entire relevant Q-channel information must be received, over a number of multiframes, for the PT to move into the idle locked state. The Q-channel information allows the portable to adjust its behavior to match the timing and the operating capabilities of the FP.

In the idle locked state, the PT may sleep for most of a multiframe, but periodically it must wake up and look in frame 0 for paging information. There are modes of operation in which the PT wakes up more or less often; however, those modes are not required in a GAP terminal. Occasionally in the idle locked state, the PP voluntarily may rescan for RFPs having a higher signal strength; if it finds one, it may relock to the new RFP. There also are certain conditions, such as timeouts in receiving information from the FT that require the PT to rescan and resynchronize with an FT.

5.5.6 Paging

Paging a handset involves sending identical paging messages simultaneously on all the beacons in an entire MAC cluster. The paging request from the DLC layer, a MAC_PAGE-req primitive, specifies one of two possible paging types: the normal paging procedure or the fast paging procedure. Only the normal paging procedure is used by GAP-compliant systems.

Occasionally, the MAC layer is asked to issue a zero-length paging message. Zero-length paging messages contain no B_S-channel data (i.e., no paging identities) and therefore do not alert any specific portable. They just contain MAC-layer information on such matters as radio blind slots, which is useful to locked portables to understand how to respond to paging messages. At most, one zero-length paging message is allowed per multiframe, but if there is no paging information to be sent, a zero-length paging message has to be sent at least every 10 seconds, replacing the low-priority N-channel information otherwise carried in frame 0 in the absence of valid P-channel messages.

When a real paging request is received, the MAC layer allocates it a lifetime of 16 multiframes before it is considered stale and gets discarded. During that time, assuming capacity allows, the paging message is broadcast at least once and perhaps up to three times. Experience suggests that multiple attempts often are necessary in real radio systems.

Under the normal paging process (not the fast paging process), a pending paging message is sent in frame 0 of a multiframe. If there are still more valid paging messages awaiting broadcast, the frame-0 P-channel message has its *extend bit* set, and more P-channel data are sent in frame 2. To meet demand, this method can be extended to provide P-channel data in up to six frames of the multiframe (all even frames but one). One normal paging message fits into one 10-ms frame. That means the raw paging capacity for a single MAC cluster is 37.5 pages per second. However, if a paging message on average has to be sent twice, the actual capacity will be half the raw capacity.

The paging procedure means that locked portables may sleep for an entire multiframe and wake up only for frame 0 to see if there are any paging messages. This is their normal idle locked mode. When they do receive a paging message, the B_S-channel data is passed up to the DLC layer for checking. If there is a lot of paging, extending across the entire multiframe, portables may have to stay on longer listening for paging messages, and their power consumption will increase.

There is a fast paging process, in which paging data can be sent more quickly to portables (such as data terminals) requiring fast call setup at the expense of increased power consumption. In that case, a paging message is sent in the first permitted frame within the multiframe. However, portables must be aware of this or they will miss messages if they wake up only for paging messages in frame 0. GAP portables do not have to do that. The fixed system also may allow portables to enter low-duty cycle mode if it sets one of the bits in its FP capabilities in one of the Q-channel messages. That allows portables to wake up every four multiframes to receive paging messages; thus, they preserve their battery capacity even longer at the expense of extra delay in setting up a call.

The identities used in paging messages can be of variable length. A short page message allows up to 20 bits of B_S-channel data to be sent in a GAP environment to address a portable by its temporary portable user identity (TPUI). However, if longer portable identities are required for other applications, P-channel messages can be extended across multiple frames.

5.6 Making and Releasing Connections

The main functions of the MAC layer are related to the setup, maintenance, and release of connections. The beacon forms a significant part of the mechanism for doing that; the individual procedures are described in this section. Many other functions of the MAC layer are omitted here, because they are related to more advanced applications, and it is left to the reader to consult the standards directly. One primary example is the set of connectionless services that can be used for the various packet data services.

Speech applications, according to the GAP standard, use connection-oriented services. Hence, call setup, maintenance, and release are described here. Furthermore, GAP requires support only for basic connection setup originated at the PT. It does not require normal connection setup, fast connection setup, physical connection setup, any connection setup originated at the FT, or call setup from an FT through a wireless relay station.

The procedures that are described here are also specific to the voice service as specified by the GAP standard. The full procedures are more general in reality, to handle various optional types of connection control. However, once you understand the GAP-specific procedures described here, you can readily appreciate general procedures from reading the MAC-layer standard.

5.6.1 Basic Connection Setup

For carrying normal speech traffic, the GAP requires what is known as a basic connection setup, which is always originated at the portable. A complete call setup is a concept of the protocols above the MAC layer, but a part of the call-setup procedure is the MAC-layer connection-setup procedure, which is described here. Formally, a connection is established when the setup procedure at the MAC layer is successfully completed, and the higher layers then are in communication and are able to finish the full call-setup process.

Within the MAC layer itself, one component of the connection-setup procedure is the bearer-setup procedure, which establishes a bearer at the PHL. Hence, within the MAC layer, the setup of a call requires one entity to do a connection setup, which in turn requires another to do a bearer setup.

5.6.1.1 Initiation of Connection Setup

As we have seen, all basic connection setups are initiated by the handset, regardless of where the call setup originated. In the case of a call origination at the handset, the NWK and DLC layers' call-setup procedures handle the user's request for a call to be set up, as part of which the DLC initiates a MAC-layer connection by issuing a MAC_CON-req primitive to the handset's MAC protocol. That initiates the procedure described next, which eventually should result in a MAC_CON-cfm primitive being returned to the DLC to indicate success.

In the case of an incoming call requiring connection setup, the fixed system's DLC layer issues a paging command to the MAC layer using the MAC_PAGE-req primitive, which results in a paging message for the target handset appearing on the FP's beacon. When the handset's higher layers receive and recognize the paging message, they initiate a connection setup just as though the handset user had requested a call setup. In this case, however, the call setup is known by the fixed system to be an answer to an incoming call and is treated as such. The incoming call setup goes only as far as creating a link that the fixed system can use to alert the user. The process is completed only after the user accepts the call.

5.6.1.2 Connection-Setup Procedure

The general procedure for a connection setup is for the DLC layer's request, the MAC_CON-req primitive, to cause the creation in the MAC layer of an MBC. It is allocated a certain amount of time to set up the connection (3s), and a certain number of attempts (11). After reaching either of those limits, it must either report success to the DLC layer or report failure and terminate.

Right at the start, the MBC may be unable to provide the requested connection (e.g., a simple GAP MAC layer is asked for an advanced connection it does not support). In that case, the MAC layer immediately issues a MAC_DIS-ind primitive in return, releases the MBC, and connection setup stops.

If the MAC layer can handle the connection, the MBC first has to ask the LLME for permission to make the call-setup attempt. The MBC may be refused permission in some cases, but then it can rerequest permission using different parameters. If, however, permission ultimately is not granted, the MBC issues a MAC_DIS-ind primitive to the DLC layer indicating the reason, the MBC terminates, and the connection-setup attempt fails.

If the LLME grants permission, the MBC attempts to create the set of bearers it needs for the requested service. In the case of a GAP system, that set contains just one duplex bearer. Upon the successful setup of the first bearer, the MBC issues a MAC_CON-cfm to the DLC layer.

At the target end, the first bearer setup request it receives will cause the LLME to be asked to set up an MBC at that end. If the request is granted and the MBC is set up, the target end's MBC will send a MAC_CON-ind to its DLC layer. The connection then is set up.

5.6.1.3 Establishment of a Single-Bearer Duplex Connection

When the MAC layer wants to set up a bearer, it first must select an available physical channel. That may come from the radio channel map it maintains, or it may choose to accept a suggested channel if it has been paged and a channel has been suggested. The MBC that wants the bearer as part of its connection makes the choice and orders a TBC function to be created. The TBC carries out the bearer-setup procedure. The primitives and messages for the entire connection setup process are shown in Figure 5.13.

To set up a bearer, the handset must wait until the base station's idle receiver is known to be listening to the correct channel. The synchronization is based on the portable's knowledge of the idle receiver scan sequence, which is obtained from broadcast system information. On the chosen carrier and timeslot, a *bearer_request* is sent. Note that bearer_request is a generic name for one of several possible messages, and it actually will be the ACCESS_REQUEST MAC layer message if the requested link is new (as shown in Figure 5.13). However, it also could be BEARER_HANDOVER_REQUEST or CONNECTION_HANDOVER_REQUEST if the new bearer is needed for a handover.

The base station must respond on the corresponding uplink timeslot (5 ms later). That may override the normal T-MUX rules. If the fixed system is

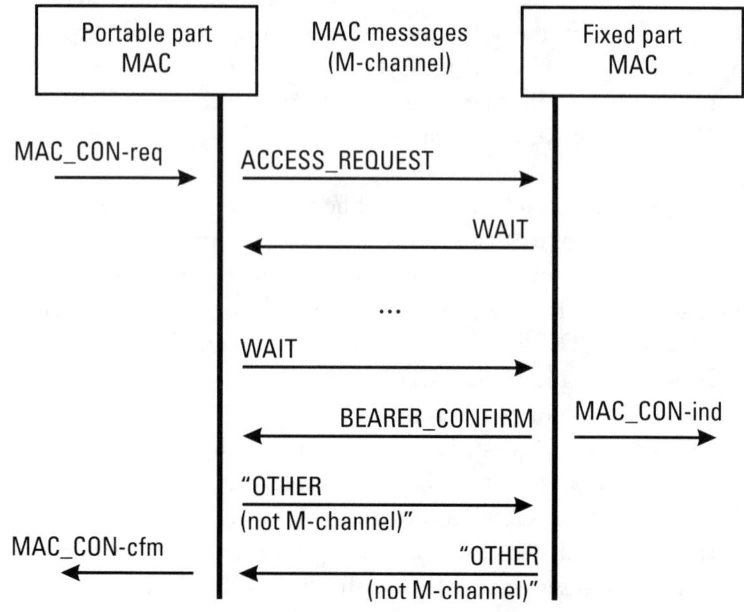

Figure 5.13 Connection- and bearer-setup procedure. *Source:* ETSI [3].

not yet ready to respond positively or negatively, a *wait* response may be sent (specifically the WAIT message), and WAIT messages will continue to be exchanged until it is ready. However, if the base station (eventually) decides to set up the bearer, it will send a *bearer_confirm* response (specifically the BEARER_CONFIRM message) and indicate to its DLC layer with a MAC_CON-ind primitive that the connection is set up. If the base station decides not to set up the bearer, it initiates the bearer release procedure.

In response to the *bearer_confirm*, the exchange of a pair of *other* messages (any A-field messages) completes the bearer-setup procedure, and the portable's TBC reports *bearer_established* to its MBC. The MBC then sends a MAC_CON-cfm primitive to its DLC layer. The higher layers (DLC and NWK) are now able to communicate for control purposes through the C_S-channel to complete the call setup.

If the bearer setup fails, the TBC reports that to the MBC with a *bearer_setup_failed*, indicating the reason, and the TBC then is released. The MBC is free to retry up to 10 more times. The reattempts are reported to the DLC layer with MAC_RES_DLC-ind primitives. If the 3s timer expires or 11 attempts fail, the MAC layer reports *setup_failure* to the DLC layer using the MAC_DIS-ind primitive and reports the event to the LLME.

5.6.2 Data Transfer and Flow Control Procedures

Once a connection has been set up, it arrives in the data transfer part of its existence.

Remember that although every available A-field in a connection is used for beacon transmissions by the MAC layer as a whole, the T-MUX ensures that space in the A-field occasionally is made available to transmit C_S-channel (control) data associated with this particular connection. Half the frames within a MAC layer–driven multiframe are allocated by the T-MUX to "M_T, C_T, N_T" priority data, that is, whenever internal MAC layer–to–MAC layer communications using the M-channel are not required, C_S-channel data may be transmitted, overriding identities information in the N-channel. The C_S-channel may contain data such as indications of keys pressed on the handset that the FP must dial.

Under the GAP, the B-field of each P32 packet is not required for any control transmissions, so it is fully allocated to the I_N_minimum_delay channel. That channel carries the user's speech communications.

The formal flow-control procedures, for all channels that support flow control, are conducted through the Q1 and Q2 bits in the A-field header (see Table 5.1). For GAP systems, only the C_S-channel uses flow control. This procedure is described next.

5.6.2.1 Slow Control-Plane and MAC Layer Internal Signaling

The size of the A-field and its structure (see Section 5.4.3 and Figure 5.8) allows only a 40-bit control field to be sent in each frame. MAC-layer messages are designed specifically to fit inside the 40-bit field, but higher layer control signaling (C_S) has to be broken up into 40-bit fragments for transmission.

Internal MAC-layer control messages are error protected using the redundancy bits of the A-field. Errors may be detected, but the receiving end just waits for the next normal transmission if a particular message is received with an error. The retransmissions are part of the MAC layer's internal procedures. However, the retransmission of C_S-channel fragments from higher DECT layers is not part of the MAC layer's internal procedures. Therefore, the MAC layer arranges for C_S-channel fragments that contain errors to be retransmitted by using what is called the MAC ARQ procedure.

Under the MAC ARQ procedure, the 40-bit control fragments are buffered and retransmitted if they are not acknowledged within a certain time. To do that, the MAC layer defines an ARQ window, which starts on the normal TDMA half-frame boundaries. In the FT, the window starts on slot 0; in the PT, it starts on slot 12. The window is one frame (10 ms) long. Only one C_S fragment may be sent in any one 10-ms ARQ window, even if there are multiple bearers for the connection. The same fragment is sent on all the connection's bearers. Because of the need for ARQ, C_S may be sent only on duplex bearers.

Acknowledgment of a downlink C_S fragment in one half-frame is given within the next half-frame if an A-field in the uplink direction is received on any bearer with good CRC and with the Q2 bit in the A-field header set to 1. For an uplink C_S fragment, the uplink acknowledgment is achieved if Q2 is 1, Q1 is 1, or both. A C_S fragment is always retransmitted until it is properly acknowledged. To do that, C_S fragments are given a 1-bit packet number, set to 1 for the first C_S fragment transmitted by an MBC, and incremented thereafter. If you look in the A-field header coding, the TA-field identifies whether the C-channel data in the tail has packet number 0 or number 1.

5.6.2.2 MAC-Layer Antenna Diversity Switching

Although antenna diversity switching is not really a flow-control procedure, it uses the same Q1 and Q2 bits in the A-field header as other flow-control procedures. Antenna diversity is specified only at the FT, and in the uplink (PT to FT) direction the Q1 bit is allocated as a request for the FT to change antennas. Note that the FT may either act on the request or not. The GAP standard does not make this procedure mandatory, and an alternative PHL procedure is often used instead (see Section 4.5.4).

5.6.2.3 Data Flow in the User-Plane

Data-flow control in the U-plane, specifically the continuous unprotected U-plane (the I_N-channel) in GAP systems, generally is limited to either passing the data on to the user or muting the data path. Each of the packets transmitted by the PHL contains one 10-ms segment of user traffic.

We have seen that there are two ways to allocate to a packet the data to be transmitted (Figure 5.9); those two ways give rise to the I_N_minimum_delay and I_N_normal_delay services. In both cases, the MBC in control of the connection asks the DLC layer for each successive I_N data segment with a MAC_CO_DTR-ind primitive, and receives the data with a MAC_CO_DATA-req primitive. In the case of the I_N_minimum_delay service, the request for data is sent just before the timeslot that is to be used, to guarantee receiving the most recent data possible. That generally requires a high level of synchronization between the MAC layer and the DLC layer. In the case of the I_N_normal_delay channel, the user data are slightly older and the timing of the request is much less critical.

In both cases, there is error control for the user data, but only in the sense that the 4-bit X-field CRC is added by the MAC layer, which may be checked by the system at the other end. If the X-field is intact, it *may* indicate good data in the B-field of the packet. However, other data also will normally be used to decide whether to pass the U-plane data unmuted to the user, including (but not limited to) the following:

- The RSSI;
- The X-field and A-field CRCs;
- The Z-field accuracy compared to the X-field;
- The presence of any S-field errors;
- The Q-bits in the A-field header;
- The measured clock jitter.

The muting algorithms for DECT and PWT systems are critical to the user's perception of the system's quality. They generally are proprietary but usually are based on the information listed above.

5.6.3 Bearer Handover

Generally, bearer handover takes place as an attempt to correct a problem, such as too many errors in the U-plane or the C-plane data traveling through the protocol stack, or when there is a low RF signal level. Section 5.6.2 dealt with flow control and listed a number of the quality indicators available to decide whether the current bearer or bearers are giving good quality service.

Bearer handover is an internal procedure within the MAC layer whereby one MAC connection is able to modify the bearers it uses while maintaining the service that the MAC layer offers the DLC layer above. Bearer handover, in effect, defines a cell cluster, in that the cluster is simply that group of cells between which bearer handover is possible.

5.6.3.1 Initiation of Bearer Handover

Bearer handover for duplex bearers may be initiated only by a PT. However, that may happen at the request of the FT, when it sends a *bearer_handover* request via a MAC-layer message. The PT may choose to ignore or accept that suggestion, and there are no formal rules for the mechanisms by which systems decide to initiate a bearer handover. For the simple reason of system stability, no more than two successful bearer handovers are permitted within a certain period defined as T202 (whose value is 3s). That allows multiple attempts at handover but means that a new bearer handover may not be initiated immediately after a successful handover.

5.6.3.2 Bearer-Handover Procedure

At the start of the bearer-handover procedure, the MBC function in charge of the connection, as shown in Figure 5.14(a) already will have knowledge of the PHL channel signal strength map (see Section 5.6.5). From the map, it selects a suitable channel and creates a new TBC. The TBC then tries to set up the requested bearer on that channel using the basic bearer-setup procedure (see Section 5.6.1 and Figure 5.13). However, instead of starting that with an

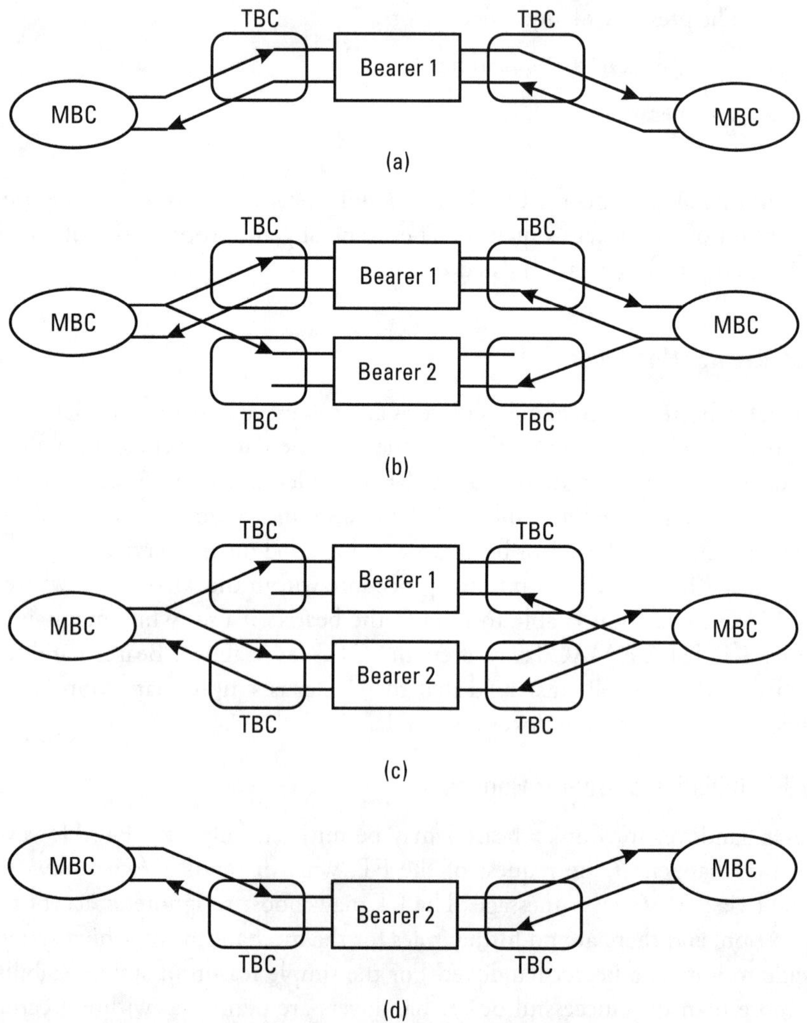

Figure 5.14 Seamless bearer handover: (a) old bearer in use; (b) new bearer set up in parallel; (c) both bearers in use; (d) old bearer released.

ACCESS_REQUEST message, it uses a BEARER_HANDOVER_REQUEST message as its *bearer_request*, to distinguish the handover from a new bearer-setup request. Because of that, both sides of the link know that a handover is being performed, even if the procedure used is the same as a bearer setup.

If the new TBC reports success to the MBC, the MBC may maintain the two bearers in parallel for, at most, T203 (16 frames) before it must release one of the two bearers (normally the older, of course). During that period, the MBC swaps over the data path to the DLC and higher layers.

In the case of a seamless handover, the MBCs at both ends first of all have to arrange for their traffic to be sent in parallel on both bearers, as shown in Figure 5.14(b). Hence, each MBC uses two MAC_CO_DTR-ind primitives per frame to retrieve the user traffic from the higher layers. The use of the I_N_minimum_delay channel is essential to achieve low delay for the speech connection, as it always retrieves the latest user data, instead of retrieving data frozen on the last frame boundary (see Section 5.4.2.1 and Figure 5.9). However, that requires the MAC and DLC layers to be highly synchronized throughout a complete system to achieve seamlessness (see [1] Annex H).

After both MBCs are sending user traffic on the new bearer as well as on the old one, they both may take their user traffic from the new bearer without a break, as shown in Figure 5.14(c). At that point, they both may drop the old bearer without the user being aware of anything, as shown in Figure 5.14(d). The whole seamless handover process from transmitted speech to received speech via the air interface is illustrated in Figure 5.15.

5.6.4 Connection Release

Connection release is the last of the three phases of a connection-oriented MAC service (setup, data transfer, release). The procedure releases an existing MBC and its bearers and reports that to the LLME and DLC layer (if required).

5.6.4.1 Initiation of Connection Release

Several events may initiate connection release:

1. The DLC of either side requests connection release with a MAC_DIS-req primitive.
2. The MAC layer has just created an MBC in response to a connection setup request, but the MBC cannot provide the requested service.
3. After a bearer release has occurred, the released TBC reports *connection_release* as the reason, because the bearer release was initiated by the other end requesting a connection release.

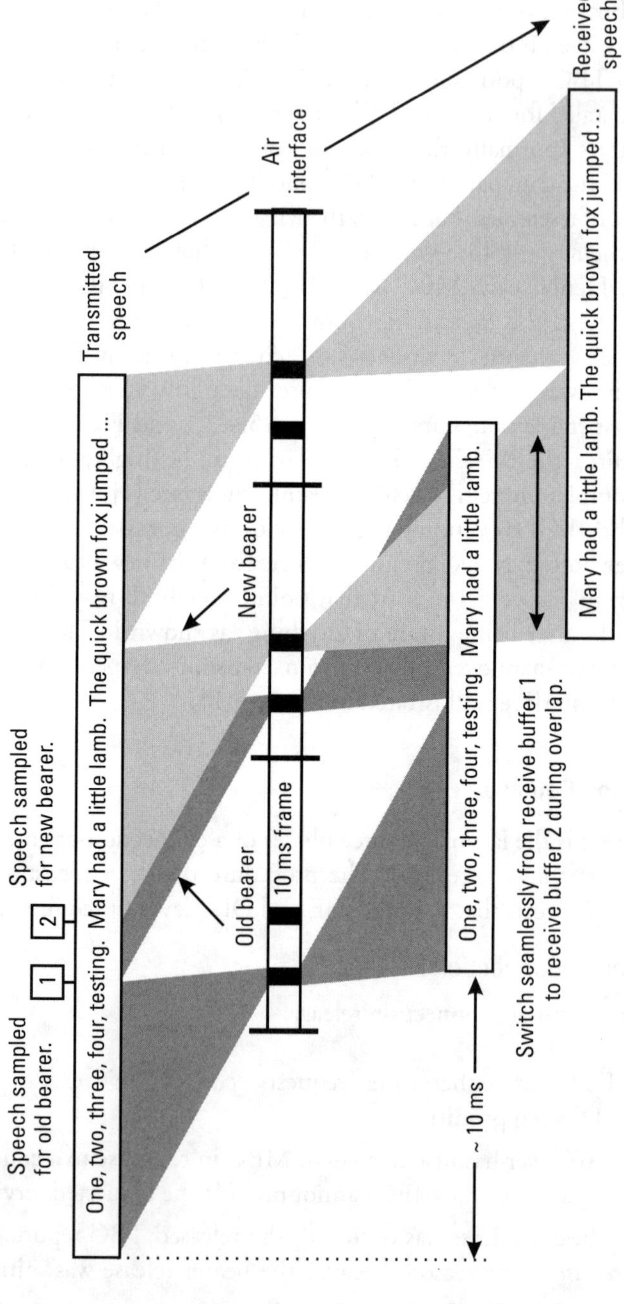

Figure 5.15 Seamless traffic handover.

4. The last TBC belonging to the MBC is released for any reason.
5. Due to bearer release for any reason, the MBC cannot maintain the required minimum connection bandwidth.

Event 1 is the usual initiation of the release procedure. Under normal circumstances, events 3, 4, or 5 will soon happen at the other side of the connection. Events 4 and 5 can happen at any time, due to unexpected bearer failure.

5.6.4.2 Connection and Bearer Release Procedure

In response to a connection release request from the DLC layer, the MBC in charge of the connection initiates a bearer release at each of its TBCs and disconnects them all. The MAC layer then releases the MBC and reports that to the LLME. The actual bearer-release procedure may be acknowledged or unacknowledged. The procedure is acknowledged only to release double simplex bearers from the receiving side; nonacknowledgment (Figure 5.16) applies to other bearers, including the GAP duplex bearer.

The RELEASE message is sent twice without prior warning, replacing normal transmissions. It is sent on the bearer to be released in successive frames, after which the transmitting end releases the radio channel immediately, without waiting for acknowledgment. At the other side of the connection, the TBC receives the message and releases its bearer. Because that is done without orders from its own MBC, the TBC reports it to the MBC, indicating *connection_release* as the reason (event 3). The MBC then releases the TBC.

5.6.5 Channel Monitoring and Dynamic Channel Allocation

Prior to first transmissions on any bearer, a DECT PT or FT has to select the bearer according to certain rules based on the measured signal strength on the physical channels to be used by that bearer. The rules depend on which type of bearer is to be used, and the PWT rules differ significantly in detail from

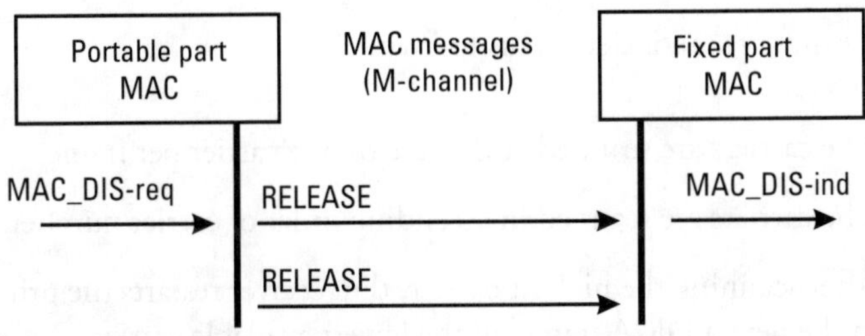

Figure 5.16 Connection release. *Source:* ETSI [3].

those of DECT. That is because of the requirement for PWT to share a frequency band with other systems, and the additional rules arise from the U.S. FCC spectrum-sharing etiquette [5].

FPs and PPs have to maintain an up-to-date map of the signal strengths on most of the physical channels they can use. Generally, measurements older than 30s are not regarded as current. Systems must resolve differences in signal strength of 6 dB or lower, from a specified minimum level that differs between DECT and PWT.

Channels where the signal strength is below the specified minimum may be considered quiet and may be immediately selected for use without making a complete list of all channels. The system also may define its own upper limit for signal strength above which a channel is considered blocked and therefore unavailable for use.

Channels with a signal strength between the lower and upper limits are candidates for selection. However, certain of those channels may be marked as unusable. For example, a system's own blind slots may be excluded (see Section 4.5.5), where its synthesizer cannot use some channels simply because it cannot change frequencies fast enough. The blind slots of the target system taken from a beacon transmission, channels on unsupported frequencies, and channels already in use also may be excluded.

5.6.6 Idle Receiver Control

Idle receivers at an RFP are not preferred. If a receiver in an RFP is not active, that is, not currently receiving a traffic bearer, it normally is not left idle at all but is set to scan the physical channels listening for bearer-setup requests. It is the IRC entity that controls that action.

Whenever an RFP has a spare receiver, that receiver is assigned to perform the primary scan, which is a defined sequence in which a receiver scans all the available physical channels. At all RFCs within an internal handover area, the primary scans have to be synchronized to scan the same RF carrier at the same time.

The primary scan is defined as follows:

- The carriers are scanned at the rate of one carrier per frame.
- The carriers are scanned in ascending order of carrier number.
- After scanning the highest carrier, the receiver restarts the primary scan in the next TDMA frame on the lowest available carrier.

If a second receiver is free, it is assigned to the secondary scan sequence, which lags behind the primary scan by six TDMA frames. If there is a third free receiver, it is assigned to the tertiary scan, which lags three TDMA frames behind the primary scan. A portable can get enough information from an FP's beacon to fully predict on what channels a bearer-setup attempt might succeed.

5.6.7 MAC Layer Identities

The subject of system identities is covered in Section 8.4. Look there for information about how a portable system manufacturer or installer first obtains and assigns a PARI (primary access rights identity) for a cordless system, how SARIs (secondary ARIs) are assigned, and how a RPN (RFP number) is assigned to each cell in a system. The assigned PARI and RPN are made available to the FT MAC layer by the LLME, and the FT MAC layer converts them into the RFP identity that is broadcast in the N-channel identities message. The LLME also makes available the assigned SARIs that are broadcast in the Q-channel of the beacon.

The MAC layer itself assigns and uses two main identities: the PMID (portable MAC identity) and the FMID (fixed MAC identity). You will see these used in the connection control messages (see Section 5.4.5.4) as means to identify the two ends of a link. The FP MAC layer derives the FMID from the RFPI's least significant 12 bits. It is not a unique identity, but just different enough to avoid co-channel interference in the initial phase of call setup, since the lowest bits are the RPN which identifies locally a particular cell. In the portable, the PMID is usually set equal to the TPUI, which is the identity assigned to a portable by the memory management procedures in the higher layers of the DECT protocol. It is the identity that is then used for paging the portable.

Finally, the DLC layer assigns a MAC connection end-point identity (MCEI) to each connection it requests. It does that because there may be many connections within a MAC layer, and the DLC layer has to make sure the primitives it exchanges with the MAC layer are from or for the correct connection. The values of MCEIs are a local matter between the MAC layer, the DLC layer, and the LLME.

References

1. ETSI, *Digital Enhanced Cordless Telecommunications (DECT); Common Interface (CI); Part 3: Medium Access Control (MAC layer) Layer*, ETS 300 175-3, Sophia Antipolis, France, September 1996.

2. TIA/EIA, *Personal Wireless Telecommunications Interoperability Standard (PWT); Part 3: Medium Access Control (MAC layer) Layer*, TIA/EIA 662-3-1997.

3. ETSI, *Digital Enhanced Cordless Telecommunications (DECT); Generic Access Profile (GAP)*, EN 300 444, Sophia Antipolis, France, August 1997.

4. TIA/EIA, *Personal Wireless Telecommunications Interoperability Standard (PWT); Part 9: Customer Premises Access Profile (CPAP)*, TIA/EIA 662-9-1997.

5. Federal Communications Commission, *Unlicensed Personal Communications Service Devices*, CFR 47 part 15, subpart D.

6

The Data Link Control Layer

6.1 Introduction

The DLC layer is rather a boring part of the DECT/PWT protocol. It can be somewhat obscure unless you are a data communications expert, but it is necessary nevertheless. The DLC layer is concerned mainly with worthy causes, such as providing increased communication reliability. As you have seen in Chapter 5, the MAC layer attempts to remove the radio system's errors in some logical channels using an ARQ protocol (notably the slow control channel). The DLC layer provides further error protection for the control channels and for some of the user channels.

Within the hierarchy of concepts that connect a telephone call to the PHL's physical channels (see Figure 5.2), the DLC layer creates links from the MAC's connections; if asked to, it also can protect some types of links from the effects of the unreliable radio channel. The links are then offered to the NWK layer as services for it to use. In providing a link, the DLC layer also adapts the structure of the data it exchanges with the NWK layer to fit into the data structures exchanged with the MAC layer (Figure 6.1).

At the top of the DLC layer, control messages and user traffic are exchanged with the NWK layer. It will be seen in Chapter 7 that the U-plane of the NWK layer does nothing but relay user data, so in reality the data can be exchanged directly between the DLC and the application via its interworking unit. The data structures at the top of the DLC are closely aligned with the telecommunications service that the equipment is delivering, such as a speech connection or a packet data service.

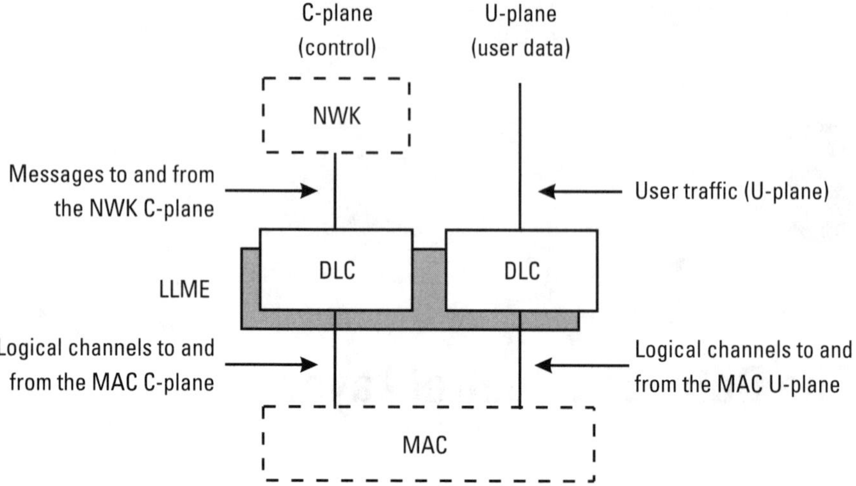

Figure 6.1 The DLC layer relative to the other protocol entities.

At the bottom of the DLC layer, logical channels are exchanged with the MAC (see Section 5.4.2), each in a sequence of *fragments*, whose size is defined by the requirements of the MAC multiplexing and the particular PHL data packets to be used.

In spite of the remarkably long DLC layer standard [1], not much of it applies to equipment providing just speech service. Speech systems normally have to conform to the GAP standard [2] or the CPAP standard [3], and, just like the MAC and the NWK layers, the DLC layer is a menu of features, not all of which are required. If you take a copy of the GAP or CPAP and mark up the DLC layer standard in one color where the profile refers to various clauses and another color elsewhere, you will see that very little of the DLC layer applies. Remember, however, that certain sections of the DLC layer are not referenced explicitly by the profile but are required implicitly or explicitly by those parts that are referenced. In this part of the book we are concentrating on basic systems that provide speech service, so we concentrate on those parts of the DLC layer required by the GAP and the CPAP.

6.2 Organization and Basic Concepts

Four quadrants are defined in the structure of the DLC layer. They are created by dividing the DLC layer into halves, the C-plane and the U-plane, which deal with control signaling and user data transport, respectively. Then the

DLC layer is further divided into the upper DLC layer and the lower DLC layer (Figure 6.2). The upper DLC layer communicates with the NWK layer, and the lower DLC layer communicates with the MAC layer. Each quadrant may contain a DLC-layer protocol entity assigned to a specific task. The DLC layer's LLME coordinates the actions of the whole DLC layer.

A single DLC-layer implementation is permitted to span more than one MAC-layer cluster of cells. Remember that at the top level of the MAC layer there are the cluster control functions and that the cluster is the entity within which the MAC layer handles bearer handover. The cluster comprises a maximum of 255 cells. The alternative type of handover, connection handover, is handled by the DLC layer. Because the DLC layer can span many clusters, connection handover, although a more complex process, effectively can handle handover within much larger cordless systems. It also acts as a backup when the MAC connection fails for any reason.

6.2.1 Messages, User Data, Frames, and Fragments

At the top of the DLC layer, communication with the NWK layer takes place through the use of primitives, which are described in the DLC-layer standard. As with all primitives in the DECT and PWT standards, the primitives are not mandatory and are used just as models in the standard for the communication

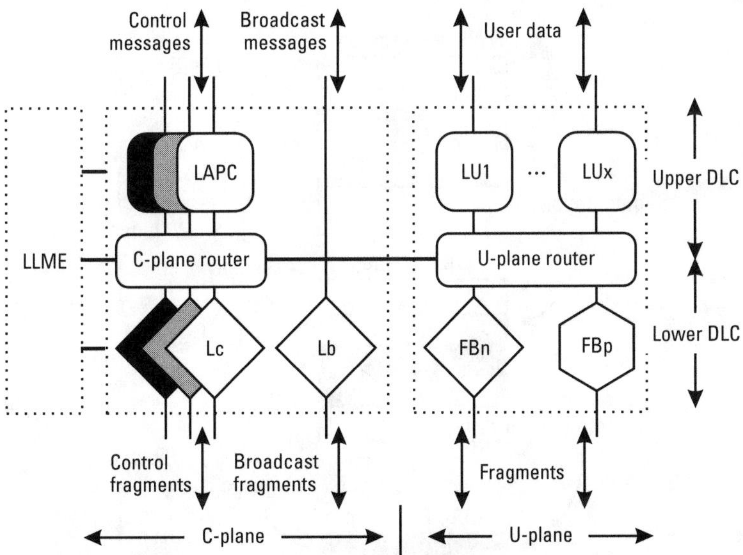

Figure 6.2 Reference model of the DLC layer. *Source:* ETSI [1].

that has to take place and the information that must flow. They can be used or not used in a real implementation. These primitives carry either control/broadcast data (the C-plane) or user data such as speech (the U-plane).

In the middle of the DLC layer, between its upper and lower halves, the DLC layer deals with frames of data. The DLC-layer standard defines frame structures, each suited to a particular data type exchanged with the NWK layer. For example, as you will see later, user data carrying speech is merely arranged, unaltered, into frames of 40 octets (320 bits) according to certain simple rules. No sequencing or error-protection information is added, and the frames are just passed on to the lower DLC layer. The 320-bit frame fits nicely into the payload section (the B-field) of a P32 PHL data packet.

In the case of control information (i.e., the NWK-layer messages for controlling a connection-oriented service such as a telephone call), the upper DLC layer needs to ensure that whatever it receives from the NWK layer reaches the DLC layer and NWK layers at the other end of the link in the right order and without error. The DLC layer first breaks up long messages, if required, into suitably sized segments, a process called *segmentation* (Figure 6.3). Next, it creates a frame from each segment by adding various octets of error-control and numbering information. That allows the receiving DLC layer to detect the need to ask for a retransmission of a frame that has errors (or that has not been received at all). In the receive direction, the reverse process of extracting

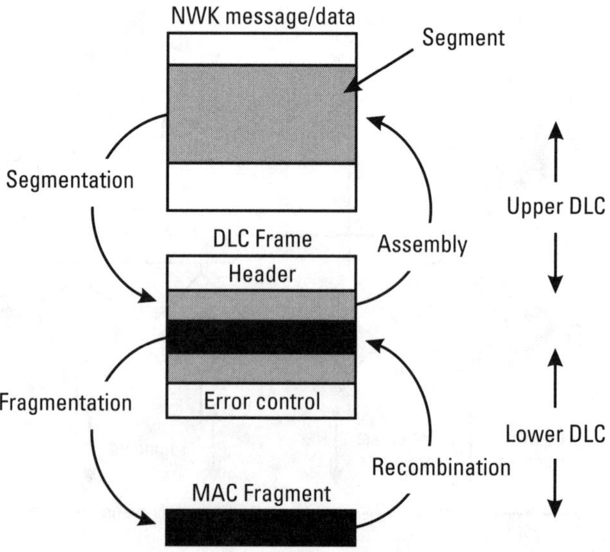

Figure 6.3 Data-structuring processes in the DLC layer.

message segments from frames and putting the messages back together is called *assembly.*

Finally, at the bottom of the DLC layer, the MAC layer uses data in fragments of a size that it normally can fit into a single data packet at the PHL. In both the U-plane and the C-plane, the lower DLC-layer entities receive frames from the upper DLC layer and perform *fragmentation,* to convert the frames into fragments. In the receive direction, they perform a process called *recombination* on the fragments they receive before passing complete frames to the upper DLC layer.

With a speech channel, as is the case in the upper DLC layer, the lower DLC layer's process is simple. That is because the fragment size and the frame size both are defined as 40 octets, based on the capacity of the P32 packet's B-field. In the control channels, however, frames are allowed to be up to 72 octets long, designed to accommodate most of the NWK layer's messages plus the control and numbering information in the upper DLC layer. The fragments destined for the slow control channel are defined as 5 octets long, to match the 40-bit capacity of the P32 packet's A-field. Hence there is a nontrivial job of fragmentation and recombination to do in the C-plane.

6.2.2 The Lower DLC Layer

The lower DLC layer has to provide information to the MAC layer and receive information from it. There are several logical channels, each of which has to be provided to the MAC layer in fragments of a particular number of octets, depending on the channel and how the MAC layer handles the information (Table 6.1). There are, therefore, slightly different fragmentation and recombination functions in each of the lower DLC layer's entities, depending on the specific data being carried and the most relevant frame size. Note that in the case of the B_S-channel, packet P00 is used for broadcasting the beacon when there is no other traffic, and packet P32 is used to combine the beacon

Table 6.1
Communication of GAP/CPAP Logical Channels With the MAC Layer

Channel	Information	PHL Packet(s)	TDMA Frame Capacity
B_S	Broadcast (paging)	P00 and P32	20 bits
C_S	Slow control	P32	40 bits
I_N	Unprotected speech	P32	320 bits

with speech traffic. Also, the B_S data are only 20 bits per frame because the GAP paging procedure adds a further 20 bits of MAC-layer data to fill up the 40-bit signaling capacity of both packets' A-fields.

The MAC may impose those limits based on its structure and the capacity of the PHL packets it uses, but as far as the user of the system is concerned, the limits are artificial and have no bearing on the service being provided. That is why it is the function of the lower DLC layer to adapt the MAC layer's data fragments to and from the DLC layer's frames, each of which is matched by the upper DLC layer to a specific user-oriented service that the DLC layer handles.

Thus, the basic transmission function of the lower DLC layer, whether C-plane or U-plane, is to fragment the DLC-layer data frames it receives from the upper DLC layer, buffer the fragments, and pass them on as the MAC layer requests them. The basic reception function of the lower DLC layer, then, is to receive fragments of data from the MAC layer when they are made available and to recombine them into frames of data that it can present to the upper DLC layer.

In the lower DLC layer, a protocol entity called Lc is used to fragment and recombine C-plane control data. In GAP or CPAP systems, Lc only has to carry connection-oriented slow control data (C_S). In more general systems, it also may have to handle fast connection-oriented signaling (C_F) and connectionless signaling data (CL_S, CL_F, or G_F).

A lower-DLC entity called Lb deals with paging messages in the C-plane, which alert a handset that it has an incoming call. From those messages, Lb creates the slow broadcast channel in the MAC (B_S).

In the U-plane, unprotected user data are dealt with by the FB_N entity, which may generally deal with both connection-oriented data (the I_N-channel) and connectionless data (the SI_N-channel). However, FB_N only needs to carry I_N for speech services. U-plane error-protected user data (I_P-channel and SI_P-channel) pass through FB_P. Those channels typically are used for data services, and neither of them is required in voice systems.

6.2.3 The Upper DLC Layer

In the upper DLC layer, telecommunication-oriented data for transmission is segmented into frames that the lower DLC layer can then deal with. In the receive direction, frames are assembled, as required, into complete messages or other service-specific data structures.

The C-plane entity in the upper DLC layer is called LAPC. It is the most complex DLC entity and deals only with control signaling and does not handle broadcast data. Note that no upper DLC entity handles broadcast data; the

NWK layer hands paging messages to the DLC layer that are already in a form suitable for the lower DLC entity Lb to fragment. Technically, like all upper DLC entities, LAPC segments NWK-layer control messages into frames. However, messages that are short enough (no more than 63 octets) will fit into one 72-octect DLC layer frame, and the NWK layer in GAP equipment is not allowed to generate messages longer than 63 octets. That means segmentation is not required. In its full generality, LAPC actually uses one of three classes of operation called U, A, and B (Table 6.2), depending on the requirements of the current service.

In the U-plane of the upper DLC layer, the protocol entities are matched to the specific telecommunications service being carried. Those entities are called LU1, LU2, LU3, and so on; their services are listed in Table 6.3. The family members missing from the table, LU8 to LU15, are not yet defined.

Because the GAP requires only one of those entities (LU1), the upper DLC U-plane is simple in speech systems. LU1 adds no control or error-protection information to the speech data it receives and just passes the frames down to the FB_N entity in the lower DLC layer for transmission. Such simplicity is deliberate. The protocol was designed to have as simple as possible a data path for normal speech service. Because the NWK layer above the DLC layer is empty in the U-plane, there is in practice a direct path for 320-bit frames containing speech from the application to the MAC layer, where they get assembled into P32 packets and sent over the radio interface.

Table 6.2
LAPC Frame Types

Class	GAP/CPAP	Function
U	No	Control messages not requiring retransmission on error. For control of connectionless communication such as point-to-multipoint packet data systems.
A	Yes	Control messages requiring retransmission on error, where there will be at most one unreceived frame at any one time. For control of simple connection-oriented point-to-point services, such as speech systems.
B	No	Control messages requiring retransmission on error, where there may be several unreceived frames at any one time. For control of more complex connection-oriented point-to-point services where multiple connections may exist between the link ends.

Table 6.3
The Upper DLC Layer's Telecommunications Services

Entity	GAP/CPAP	Telecommunication Service
LU1	Yes	The transparent unprotected (TRUP) service; used for speech, without additional error protection.
LU2	No	The frame relay service.
LU3	No	The frame switching service; not fully defined.
LU4	No	The forward error-correction service; not fully defined.
LU5	No	The basic rate adaptation service. It provides 8, 16, 32, and 64 kbps data channels.
LU6	No	The secondary rate adaptation service. It builds on LU5 to provide ITU-T V-series asynchronous data rates from 50 bps to 19,200 bps. Also provides synchronous data transfer at up to 56 kbps.
LU7	No	A 64-kbps bearer service with enhanced error protection.
LU16	No	A nonstandard (proprietary) service.

6.2.4 Routing

Between the upper and lower DLC entities, in both the C-plane and the U-plane, is a *router.* In the simplest system, a GAP-compliant cordless telephone supporting just one call at a time, the control signaling channel uses just one instance of LAPC and just one Lc. The routing could hardly be simpler: the entities simply talk to each other, exchanging frames of data. Similarly, in the U-plane, there is just one LU1 entity in the upper DLC layer and just one FB_N entity in the lower DLC layer. Again, the term *routing* is an exaggeration for the direct communication that takes place.

While in simple fixed equipment there also is only one connection to handle at a time, in a more complex fixed system (e.g., a residential set with two lines or a PABX), there may well be more than one simultaneous connection. Each will have its own pair of LAPC plus Lc entities for control signaling and LU1 plus FB_N pairs for the speech path. That is where the router is needed. It is the job of the LLME to ensure that the correct connections are established through the router between the appropriate upper and lower DLC entities. Those connections must be maintained throughout the connection and also need to be maintained after a connection handover.

6.2.5 Identities

Identities are used in the DLC layer to identify the correct MAC connection for a particular DLC-layer connection to use and to identify the correct entity with which to communicate in the NWK layer. In simple systems supporting a single connection, there may be no alternatives. Above the level of a one-line residential telephone set, however, there may be many simultaneous connections; therefore, identities are required.

The DLC layer is handed most of the identities it needs at link-setup time. At the originating end, those identities come from the NWK layer, associated with the request to set up the link. At the target end, the information is received from the originator via the MAC layer. At connection setup, the NWK layer hands the DLC layer a piece of information called the data link endpoint identifier (DLEI). The DLC standard provides a description of how to extract from that identifier all but one of the relevant identities it needs: the SAPI (service access point identifier), DLI (data link identifier), LLN (logical link number), MCI (MAC connection identifier), ARI (access rights identity), and PMID (portable part MAC identity). Details can be found in the standard and are not reproduced here.

The one remaining identity used by the DLC layer is called the MAC connection endpoint identity (MCEI). The MCEI is used only at the interface between the DLC layer and the MAC layer to make sure the DLC layer is in communication with the correct MAC connection. Since it is used only for that purpose, it is left as a local matter for the MAC/DLC layer implementer to choose any suitable identifier to allocate and use, as long as the MAC layer and the DLC layer agree.

6.3 Data Flow Through the Lower DLC Layer

There are four different protocol entities in the lower half of the DLC layer: Lc, Lb, FB_N, and FB_P. All those entities have a generic function but with special features depending on the type of data they are carrying. In general, the lower-DLC entities all store frames of data received from the upper DLC layer and provide them in sequence to the MAC layer in a series of fragments. In the receive direction, they receive and store fragments from the MAC layer and assemble them into frames for the upper DLC layer to handle. The four protocol entities each handle different logical channels at their interfaces to the MAC layer (Table 6.4), but the voice system profiles require support for only some of them.

Table 6.4
The Lower DLC Layer's Protocol Entities

Entity	Channel	GAP/CPAP	Use
L_C	C_S	Yes	Slow control for connection-oriented links
L_C	C_F	No	Fast control for connection-oriented links
L_C	CL_F	No	Fast control for connectionless links
L_C	CL_S	No	Slow control for connectionless links
L_C	G_F	No	User control for connectionless links
L_B	B_S	Yes	Broadcasting (typically paging)
FB_N	I_N	Yes	Unprotected connection-oriented data
FB_N	SI_N	No	Unprotected connectionless data
FB_P	I_P	No	Protected connection-oriented data
FB_P	SI_P	No	Protected connectionless data

6.3.1 The Lc C-Plane Entity

Whenever a cordless link is established, it needs the services of one Lc entity to conduct communication of control data with the MAC layer. The request from the NWK layer at the originating side triggers the DLC layer's LLME into creating an Lc entity for the link. The LLME then issues a request to the MAC layer to create a connection. At the target end of the link, it is the establishment of the MAC connection and the notification from the MAC layer that causes the target DLC layer's LLME to establish a peer Lc entity. In both cases, the setup procedure ensures that the Lc entity knows the identity of the particular MAC connection with which it is associated. Because of that, it can correctly communicate with a MAC layer that may be controlling more than one connection.

The Lc entity handles a number of functions associated with the control signaling for a connection:

- Routing of frames to logical channels;
- Channel-dependent fragmentation on transmit and recombination on receive;
- Checksum generation on transmit and checking on receive;
- Identification of the correct frame boundaries on receive;
- Connection handover.

6.3.1.1 Routing of Frames to Logical Channels

Normally, the LLME instructs the Lc entity which logical channel to use for sending each frame it receives. However, the GAP and CPAP standards demand only that Lc support C_S, the slow control signaling channel for connection-oriented links. Nevertheless, a DLC layer capable of handling more than just GAP speech connections has to make that choice for each frame and assign a logical channel for the frame.

During each TDMA frame, the MAC indicates whether it is ready to transmit the next fragment from the DLC layer by sending a MAC_CO_DTR-ind primitive, indicating the maximum number of fragments it wants. For the C_S-channel, that is a maximum of one, although some other channels can handle more than one fragment per TDMA frame. It also could be zero if the MAC layer has to retransmit a previous fragment using the MAC ARQ procedure, and it does not want to receive another one during this TDMA frame. The DLC layer responds with a MAC_CO_DATA-req containing the next fragment, if there is one, and then deletes the fragment from its buffer. Whenever a fragment is received, the MAC layer hands it to the DLC layer in a MAC_CO_DATA-ind primitive.

The process takes place when the MAC connection is in the open state, that is, after it has been asked to open a connection with a MAC_CON-req and has confirmed with a MAC_CON-cfm primitive. The DLC layer is permitted, at risk, to send fragments to the MAC layer after the first of those primitives but before the second, while the MAC is still in the open pending state. That reduces the call-setup time, but in this case, the DLC layer must keep the data it has sent until the second primitive is received and must resend it if the opening of a MAC connection fails the first time.

6.3.1.2 Checksum, Frame Boundaries, and Recombination

Lc does not know anything about the content of the frame it has to transmit other than its length. It just receives a frame of a specified length from LAPC and has the job of buffering the frame, computing a checksum, inserting the checksum into the last two octets of the frame, fragmenting the frame, and sending the fragments in the C_S-channel until the MAC layer has requested and accepted them all.

On receive, the DLC layer's Lc does not know in advance how many fragments are in the frame that is coming in from the MAC layer. Because it has to find out where the frame boundaries lie, on receiving a fragment, Lc adds it to its buffer and checks to see if it can identify a frame. Because of the MAC ARQ, Lc may assume initially that the data it receives are error free. Thus, it can move a pointer to the start of a candidate frame along the buffer and determine for each possible frame a length from the frame's header (covered later, in Section 6.4.1.1). From the length, Lc can determine the position of a candidate checksum and see if it is correct for the frame. When it finds a correct checksum, the Lc entity is in DLC-frame synchronization. Note, however, that any error that escapes the MAC ARQ's error protection may cause Lc to have to reestablish DLC-frame synchronization.

6.3.1.3 Fragmentation

A complete frame of slow control channel data, received by Lc from LAPC in the upper DLC layer will already be quantized into a multiple of 5 octets so that Lc can send it in 5-octet fragments via the C_S-channel. An Lc entity set up to handle the fast control channel, C_F, would expect the quantization to be in 8-octet steps.

A slow control channel frame received from the upper DLC layer may be any size from 5 to 70 octets (from 8 to 72 for C_F). It contains a header with link-control parameters and space for a 2-octet checksum, leaving room for a segment of up to 63 octets of control signaling from the NWK layer. The control segment might be a complete NWK layer message, or it might be part of a longer message. The lower DLC layer does not care about that—it handles only one frame at a time in the order it receives the frames.

Lc always aligns the fragments of a frame it sends down to the MAC with the TDMA frame. In other words, Lc performs the following functions:

- Processes frames in the order they are received;
- Breaks the frame into fragments, number 1 being the first 5 octets of the frame if it is for transmission in the C_S-channel;
- Sends the fragments to the MAC layer in ascending order.

The MAC layer is relied on to preserve the order of fragments and to retransmit them from its own internal buffering (as it does). Hence, there is no need for the Lc process to add sequence numbers for each fragment.

6.3.1.4 Connection Handover

Lc may have to take special action during connection handover. That is where a new connection is required from the MAC to carry a current service. The DLC layer at the PT will ask the MAC for a new connection in the same way it asked for the original connection, with a MAC_CON-req primitive. However, in this case, it will supply the MAC with a flag indicating that the new connection is for a connection handover, and it supplies the old MCEI. When the new connection is set up, it has the same MCEI as the old one, as indicated in the MAC_CON-cfm sent from the MAC layer to the DLC layer.

The process can happen in one of two ways. In one method, the original connection is released first and the new connection then set up. That always results in some audible break in a speech connection, but the Lc has to deal with only one connection at a time. The alternative is for the new connection to be set up in parallel with the old connection before the old connection is released. That can be used to ensure seamless handover, but the data buffering in the Lc entity has to cope with two connections in parallel for a short time, each of which, it must be assumed, will consume and provide control channel fragments at a different rate.

6.3.2 The Lb C-Plane Entity

The Lb entity handles broadcast signaling, which currently is used only for paging messages. Unlike Lc, the Lb entity exists continuously between calls and is established at system power-up. It performs the following functions:

- Buffers and forwards NWK-layer messages to and from the MAC directly, without an upper-DLC entity;
- Distributes transmitted messages over different clusters;
- Collects the received messages from different clusters.

The broadcast service, Lb, is a simple path through the DLC layer. There is no upper-DLC entity that handles broadcast data. The DLC layer at a fixed part receives paging data from the NWK layer in a DL_BROADCAST-req primitive. The data are received in a ready-framed format, of either 3-octet frames, 5-octet frames, or extended frames containing a multiple of 5 octets, up to 30 octets.

A 3-octet frame contains a 20-bit paging identity that fits directly into a short page message in the MAC. A 5-octet frame contains a 36-bit paging identity for a full paging message. The extended frames contain longer paging identities. In GAP systems, the NWK layer uses only the 3-octet frame containing the 20-bit paging identity. The DLC layer adds nothing at all to the data in the B_S-channel that it carries.

When the NWK layer asks the DLC layer to arrange for a zero-length page, no paging identity data at all flow through the DLC layer. The DLC layer merely creates the primitive to request the MAC layer to perform the zero-length page, which contains no B_S-channel data. Remember that a zero-length page message carries useful MAC-layer information for portables to use in looking out for and responding to real paging messages that indicate the arrival of incoming calls.

For all paging requests, the DL_BROADCAST-req primitive contains a list of all the MAC cell clusters to which the paging data should be sent. Lb creates separate MAC_PAGE-req primitives that contain the paging data for each cluster. In that way, the NWK layer controls how paging messages are distributed over a complete system.

Paging requests from the fixed system are received at the portable's DLC layer in MAC_PAGE-ind primitives. Even if the paging data are in error, as determined by the MAC layer from the A-field's CRC, the data are passed up to the NWK layer in a DL_BROADCAST-ind primitive. Note that no B_S-channel data flow in the receive direction whenever a zero-length page is received. However, with the arrival from the MAC layer at a portable of a zero-length page, the fact is still communicated up to the NWK layer in the primitive without any paging data. Note also that in the receiving direction, unlike Lc, Lb does not need to look for the end of a frame. That is because all paging identities of whatever length are handled by the MAC layer as complete and unfragmented units, even if the MAC layer has to send or receive them over multiple frames. That is the one exception to the general rule that fragments are dimensioned to fit into a single PHL packet. The exceptional case never arises in GAP equipment, however, since the short paging identities required by GAP always do fit into one PHL packet. Nevertheless, because of that exception, the Lb entity never has to perform recombination of fragments into frames.

6.3.3 The FB_N and FB_P U-Plane Entities

As with the broadcast entity, Lb, the two U-plane lower-DLC entities, FB_N and FB_P, are simple relays for frames of data. However, they are created only as required to handle user data during calls. FB_N handles unprotected data, for example speech, which it sends to the MAC layer via the I_N-channel. FB_P handles error-protected data (e.g., packet data), which it sends to the MAC layer via the I_P-channel and to which the MAC layer adds a checksum.

The data FB_N and FB_P transmit come from one of the LUx entities in the upper DLC layer, according to the traffic being carried. Data flow is under the control of the MAC layer, so they have to buffer data they receive from the upper DLC layer and pass the data down only at the MAC's demand. They also are required to receive frames from the MAC layer whenever a frame is offered.

The data frames that FB_N and FB_P receive from the upper DLC layer are already in suitable sizes for passing directly to the MAC and vice versa. The LU1 entity, which is used for GAP speech, already deals in 40-octet (320-bit) frames, which is precisely the size of the B-field in a P32 PHL packet. Hence, the entities do not, in this case, provide a fragmentation or recombination function. In fact, apart from the buffering function, the only other function they provide is to deal with data overflow or underflow, should there be any unwanted mismatch between the data flows into and out of the entities.

Whenever a MAC_CO_DTR-ind primitive arrives from the MAC layer, indicating that the MAC is ready to transmit a frame of data, FB_N or FB_P is required to hand the MAC layer the data it demands. The data is submitted to the MAC in a MAC_CO_DATA-req primitive. Whenever the MAC has received data to hand to FB_N or FB_P, it sends the data in a MAC_CO_DATA-ind primitive.

Because data flow is regulated by the MAC's demands, FB_N and FB_P have to either delete octets of data or repeat octets, as appropriate, to match supply and demand. This function deals with slight clock differences between the NWK layer and the MAC layer or with timing matters at connection setup, connection release, or connection handover. For most applications, clock differences are avoided, because the PT clock synchronizes to the FT clock and the FT entities and the PT entities are in internal synchronization.

6.4 Data Flow Through the Upper DLC Layer

There are two important entities in the upper DLC layer. In the U-plane, there is the LUx entity, where *x* is a value from 1 to 16. Each different LUx entity handles a different telecommunication service. For example, LU1 is the entity that handles speech. The other entity is the C-plane entity LAPC, which handles the framing and retransmission of control messages to make sure they reach their destinations in sequence and without error.

6.4.1 The LAPC Entity

LAPC exchanges control messages with the NWK layer and exchanges frames containing those messages with Lc in the lower DLC layer. It is derived from the ISDN LAPD protocol in ITU-T Recommendations Q.920 [4] and Q.921 [5]. It differs, however, in several respects and sometimes resembles more the LAPD of GSM (see GSM technical specification 04.06 [6]). LAPC provides the following functions:

- Deals with the control channel for one data link only;
- Deals with message segmentation and assembly;

- Provides flow control for the control channel;
- Deals with timeout and protocol errors, providing error recovery.

An individual LAPC entity always operates with a single Lc entity in the lower DLC layer, and the pair controls one data link via a specific MAC connection. It operates in one of three possible classes (Table 6.5), depending on whether the link provides no error recovery (class U), simple error recovery (class A), or a more comprehensive error recovery mechanism (class B). LAPC entities do not change class throughout their existence, except that a LAPC entity attempting to set up a class B link can back off to provide a class A link instead if its peer LAPC entity supports only class A operation.

GAP systems do not require support for class U or class B. Class U can be used to provide connectionless communications from a fixed system to more than one portable part. Class B can be used when there are multiple independent MAC connections between a pair of DECT systems using multiple independent MAC connections. Class B operation also offers the ability to voluntarily suspend and resume communication. Class U and class A LAPCs deal with only involuntary suspension due to, for example, connection handover or another event that unexpectedly interrupts communication.

Usually, a portable part operates just one LAPC+Lc pair, but multiple instances may appear in a fixed part that supports more than one connection simultaneously.

6.4.1.1 Framing

At the top of the DLC layer, messages are exchanged that are meaningful to the NWK layer, and the NWK layer expects to receive those messages error free. Achieving that is the job of the LAPC entity. The first thing LAPC does with a message it receives from the NWK layer is to put it into the information field of a C-plane frame of type FA. The general structure of the frame is shown in Figure 6.4(a); details of the headers are shown in Figure 6.4(b).

Table 6.5
LAPC Operating Classes

Class	GAP/CPAP	Comments
U	No	Used for unacknowledged transfer. Frames are unnumbered.
A	Yes	Used for acknowledged transfer. Frames are numbered 0 to 1. Class A is a subset of class B.
B	No	Used for acknowledged transfer. Frames are numbered 0 to 7.

The variable field sizes of this frame are characterized in Table 6.6.

The largest frame that LAPC will construct contains 63 octets of NWK-layer message; the appropriate address/control fields, and so on; and just enough padding in the fill field to bring the entire frame length (FLEN) up to a multiple of the fragment size required by the MAC layer. That length includes the 2-octet checksum field, but remember it is actually the Lc entity that inserts the checksum on transmit and checks it on receive.

6.4.1.2 The Length Indicator Field

The length indicator field in the frame encodes only the length of the information field. Entities, such as Lc, that must compute the length of an entire frame

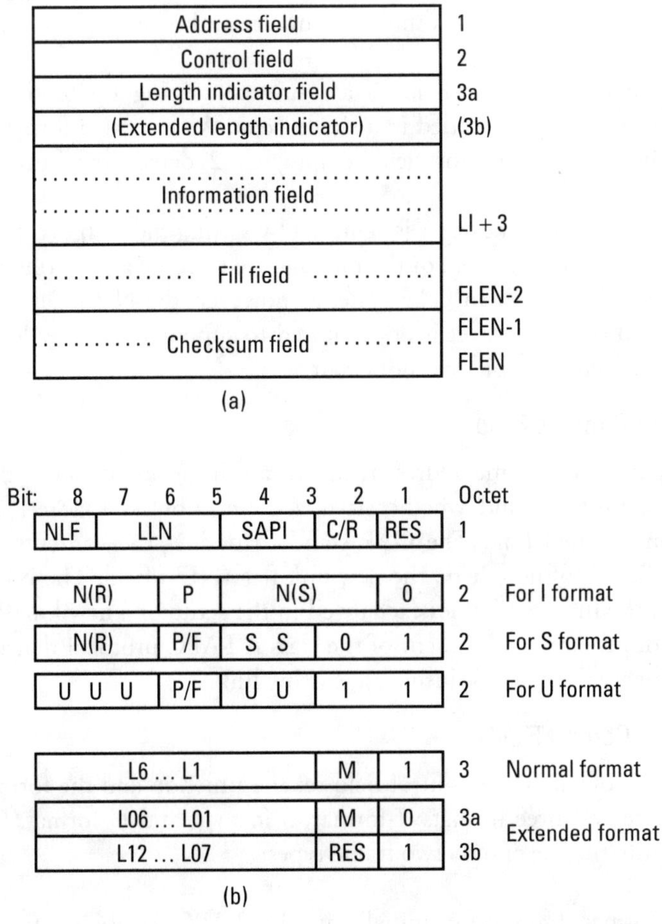

Figure 6.4 (a) The generic frame format, type FA, and (b) its headers. *Source:* ETSI [1].

Table 6.6
Characteristics of the FA-Type Frame

Parameter	C_S-channel	C_F-channel
GAP/CPAP	Yes	No
Maximum FLEN	70 octets	72 octets
Maximum NWK information field	63 octets	63 octets
FLEN quantization	5 octets	8 octets
Maximum fill field	4 octets	7 octets

must be aware of the field's complete structure and whether it contains C_S or C_F data.

The length indicator field in a frame containing up to 63 octets of NWK-layer message is encoded in a single octet. An extended format is specified for the length indicator field, employing 2 octets, but currently it is not used.

Segmentation of messages is achieved by setting the M bit to 1 whenever more frames containing more of the message are to come and setting it to 0 on the last (or only) frame. In GAP systems, however, the NWK layer does not generate (and the DLC layer is not required to support) messages longer than 63 octets, so segmentation does not occur.

6.4.1.3 The Address Field

The contents of the frame address field contain information about the operation of the protocol and whether it is a connectionless (packet) link or a connection-oriented link. The GAP standard defines the contents precisely. Only one field is of note here, the new link flag (NLF) field. The NLF field is set to 1 on the first two frames exchanged in the setup of a new link, to distinguish it from the reestablishment of the class A LAPC protocol that may happen in recovery from errors during an existing link.

6.4.1.4 The Control Field

The contents of the frame control field set the function and the format of the frame. There are three formats: I-format, S-format, and U-format. GAP and CPAP require the use of only two frame types:

- I-format frames for initializing the LAPC entity or for carrying information;

- One of the S-format messages, the *receiver ready* message, or RR response, when the LAPC entity is set up.

For the two acknowledged frame formats only (I and S), the header contains the transmitted and received frame numbers, N(S) and N(R). N(S) is the send sequence number, which is just the number of the frame. In class A operation, frames are numbered alternately 0 and 1. The N(R) field indicates that the LAPC entity has received all frames up to N(R) − 1. In class A operation, N(R) = 0 indicates that the next frame expected is a frame with N(S) = 1 and vice versa. The variables and the LAPC procedure are used to ensure that frames arrive in sequence and without error. Full details of the operation of acknowledged frame transmission and reception are contained in the DLC standard [1].

6.4.2 The LU1 (GAP Speech) Service

In the upper DLC layer, U-plane information flows through one of the LUx entities. For GAP-compliant systems carrying speech, that utility is called LU1. It provides what is called transparent unprotected (TRUP) service, which simply means that the data are relayed through the upper DLC layer in frames without any additional control, sequencing, or error-protection information.

In the other LUx entities, a substantial amount of framing, multiplexing, sequencing, and so on, may have to be done, depending on the specific service being carried. For speech, however, none of that is required and LU1 is very simple.

LU1 takes speech from the NWK layer in frames of 40 octets that fit into a single P32 packet in each TDMA frame at the PHL. The data arrive in DL_DATA-req primitives. The frames are called FU1 by the DLC layer. In general, FU1 frames can have lengths from 8 octets up to 100 octets, depending on the packet size to be used at the PHL and whether the MAC layer will protect the data from errors. The GAP speech service requires support only for the 40-octet FU1 frame.

The GAP also requires that speech is sent using minimum delay operation. In that procedure, the MAC layer asks the DLC layer for a speech frame just before the specific timeslot at the PHL in which the speech is to be sent. In the normal delay procedure, by contrast, frames of speech are sampled in alignment with the TDMA frame; therefore, the delay through transmission is longer. In minimum-delay operation, however, the speech sampling has to be done on a slot-by-slot basis. Minimum-delay operation requires good synchronization among the MAC, DLC, and NWK layers.

On connection handover, minimum-delay operation generally means that the old and new connections transmit FU1 frames and, although they carry the same speech connection, will be sampled at different times and so will contain different data. If seamless handover is required, simply swapping from one connection to the other usually means some small duplication of the speech channel as the user hears it, or a small gap. Precautions have to be taken to prevent that.

6.5 Procedures

This section covers only the basic procedures, those required in GAP-compliant equipment that provides speech. That means the setup, maintenance, and release of basic connections, plus the operation of the broadcast channel for paging messages. Advanced connections are not required. The main complexities of the protocol are handled either at the MAC layer (see Chapter 5), which is close to the radio interface, or at the NWK layer, which is close to the application. At the DLC layer in between, procedures are simpler, with the main DLC-layer complexity lying in the operation of the control signaling channel, in the LAPC entity.

6.5.1 Connection Setup

In GAP-compliant voice systems, connections are set up only from the handset. That happens in response to either a user action at the handset (i.e., going off-hook) or in response to a paging message from the base station to announce an incoming call. With one minor exception, specifically to distinguish the two different situations, the procedures are the same. Setup happens in four steps: (1) the LLME sets up the LAPC and Lc entities to support the new connection; (2) it creates the MAC connection (if a new connection is required); (3) it connects the Lc to its MAC connection; and (4) it initiates the LAPC protocol.

6.5.1.1 MAC-Connection Setup

MAC-connection setup (Figure 6.5) is initiated at the portable when the NWK layer issues a DL_ESTABLISH-req primitive containing information about how the DLC layer should set up a link. The GAP standard specifies precisely which options are required in the primitive. In summary, the primitive specifies the DLC-layer (and MAC layer) identities to be used and that the DLC layer should use a class A LAPC entity. If the establishment is happening as a result of a paging request for the handset, the primitive also will contain the first

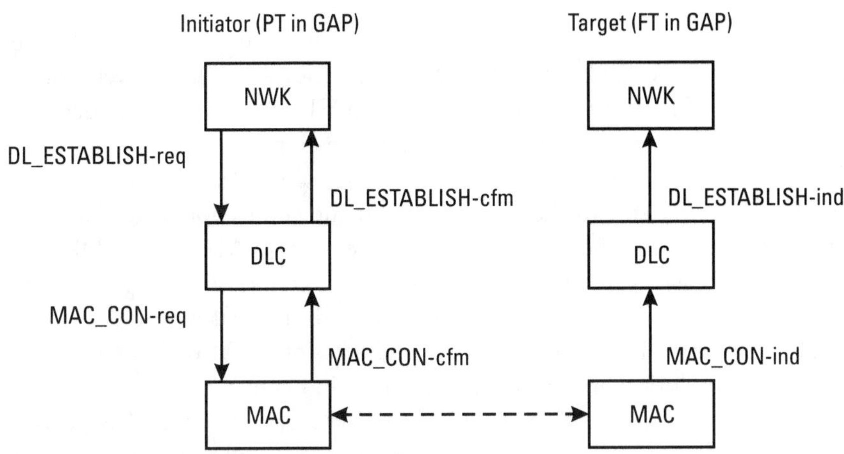

Figure 6.5 MAC-connection setup.

NWK-layer message to be sent to the base station when the connection is established, namely, a LCE_PAGE_RESPONSE (discussed in Chapter 7), which will be passed on to the base station's NWK layer to indicate that the connection is being set up in response to a paging request.

As a result of the request, the LLME creates a class A LAPC and its partner Lc. The DLC layer nominally is allowed three attempts at setting up the link, each lasting up to 2 seconds before it must report either success to the NWK layer with a DL_ESTABLISH-cfm or failure with a DL_RELEASE-ind primitive.

Whenever the connection setup request is new, the LLME decides to create a new MAC connection rather than attempt to connect the DLC layer to an existing MAC connection. Therefore, the DLC layer sends the MAC layer a MAC_CON-req primitive with parameters that do the following:

- Ask that a basic connection be established;
- Ask that the slow control channel (C_S) be used for the C-plane;
- Ask for a full slot at the PHL, for 32-kbps capacity in the U-plane;
- Ask for I_N_minimum_delay U-plane service;
- Assign a local identifier to the MAC connection for use in communication between the DLC layer and the MAC layer, that is, the MCEI;
- Assign the MAC connection a PMID, which comes from the original DL_ESTABLISH-req primitive and is used to identify the connection at the air interface.

The MAC connection setup process itself is described in Section 5.6.1. Assuming the process is successful, the MAC notifies the DLC layer with a MAC_CON-cfm primitive containing the MCEI (to allow it to be identified in systems that may have multiple MAC connections) and confirming that a basic connection has been set up. The DLC layer then returns DL_ESTABLISH-cfm to the NWK layer. If the connection setup has to be retried by the MAC, the DLC layer is notified with a MAC_RES_DLC-ind primitive for each reattempt. Then if the MAC connection setup ultimately fails, the DLC layer will receive MAC_DIS-ind and will report that to the NWK layer with a DL_RELEASE-ind. The states of the MAC layer as seen by the DLC layer during this procedure are shown in Figure 6.6.

At the target base station, MAC_CON-ind is issued by the MAC layer to the peer DLC layer if the connection is set up successfully. It contains all the above information from the portable but with a local MCEI that the fixed part will use in future to identify the new MAC connection. The target DLC layer indicates the link setup to its NWK layer with DL_ESTABLISH-ind.

6.5.1.2 LAPC Protocol Setup

Once the MAC connection has been set up, LAPC can set itself up (Figure 6.7). To do that, the LLME resets LAPC's state variables, timers, and so on, and causes the class A LAPC to send an I-frame to the peer DLC layer over the MAC connection, using a MAC_CO_DATA-req primitive. That initiates communication with the peer LAPC. The first frame has the new link flag set, but it is not set again. When the target base station receives the I-frame with

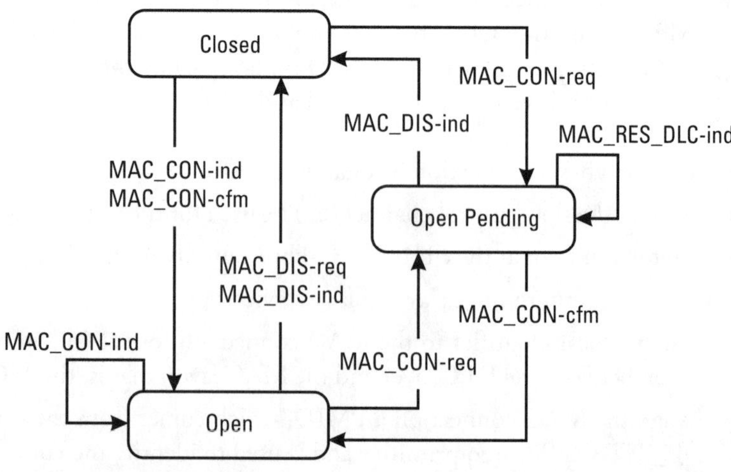

Figure 6.6 MAC states as seen by the DLC layer during connection setup. *After:* ETSI [1].

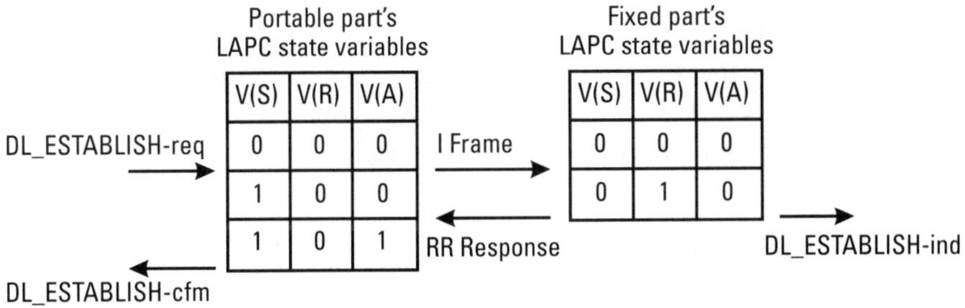

Figure 6.7 Set-up of LAPC. *Source:* ETSI [2].

NLF set, it establishes its own Lc and class A LAPC instances, resetting state variables, clearing exceptions, timers, and the retransmission counter. Then it returns an RR response frame to confirm the setup (with NLF set just this once) and informs its own NWK layer of a successful link setup with a DL_ESTABLISH-ind primitive. When the link initiator receives the RR response frame, it clears its link establishment timer and issues the DL_ESTABLISH-cfm to the NWK layer. The link is now established, and all entities are aware of it and all the identities and modes involved.

If the link establishment timer expires at the initiator before the RR response frame is received, LAPC may retransmit the I-frame and restart the timer, provided the retransmission counter is less than 3. If three attempts fail, it must issue a DL_RELEASE-ind primitive to the NWK layer and discard all buffered I-frames. The entire link setup then has failed.

Should it be necessary, it is possible for class A operation to be reestablished according to the above procedure at any time, with the new link flag cleared. All outstanding DL_DATA primitives and I-frames simply are discarded, and the state variables are reset before starting again with the transmission of the first I-frame.

6.5.1.3 U-Plane Setup

Although the preceding description may appear to apply just to the setup of the C-plane, the establishment of a connection at the MAC layer includes the U-plane.

The moment a connection is established, the MAC starts to request U-plane data from the DLC layer for transmission and provides the DLC layer with received U-plane data. In the case of speech, if the FB_N entity in the lower DLC layer already is receiving U-plane data through the NWK layer, it will be immediately connected to the MAC layer. If the U-plane still is not available via the DLC layer, FB_N has both overflow and underflow procedures that supply dummy data on transmit and discard data on receive.

This is an important area for the user's perception of the system's quality. Dummy data in the U-plane at link setup can cause unpleasant and loud sounds when decoded by the speech-processing system. Care must be taken to mute the U-plane until reliable data are being exchanged.

6.5.2 Connection Handover

In DECT and PWT systems, handover is always controlled by the portable, whether voluntary or involuntary. Voluntary connection handover may happen when the portable DLC layer initiates the procedure because of some quality problem with the underlying MAC connection, to create a new connection before it is too late. Involuntary connection handover may happen when the MAC connection fails for some reason and the portable DLC layer is forced to invoke the procedure to recover the link.

The procedure for voluntary handover is simple when the portable has decided that it is receiving poor service from the MAC connection. First, the portable establishes a new MAC connection, then it initiates normal release of the old connection.

The two processes can be performed serially or in parallel. In the case of serial handover, the old connection is released first, so inevitably there is some break in the user's service. In the case of the parallel operation, the new MAC connection is established fully before the old one is released, so there is an opportunity to achieve seamless handover without any break in the user's service. In the case of involuntary handover, the old MAC connection breaks first anyway, so the only thing the portable can do is establish a new connection as quickly as possible to minimize the break in service.

6.5.2.1 Voluntary Parallel Handover

In voluntary parallel handover, the portable DLC layer issues a MAC_CON-req to the MAC and expects to receive a MAC_CON-cfm when the new connection is established. At that point, it initiates a normal connection release for the old connection, as discussed in Section 6.5.3.

At the fixed part, the DLC layer receives an unexpected MAC_CON-ind primitive indicating the establishment of the new MAC connection while the old connection is still in place. For GAP systems, where it is not permitted to have more than one basic connection in place between two specified end points, the creation of the new connection alone is enough to inform the fixed part that this is a connection handover, that it should connect its Lc entity for this link to the new connection, and that it should release the old connection. In advanced systems where multiple connections may exist between endpoints, the DLC layer has to check the connection identities first to see if it must release a connection and, if so, which it should release.

6.5.2.2 Other Types of Handover

In the case of involuntary handover or serial voluntary handover, the fixed-part DLC layer first receives an unexpected MAC_DIS-ind indicating an abnormal disconnection. It then waits for up to 10 seconds to see if a new connection gets established. If it receives a MAC_CON-ind primitive indicating connection handover, it reconnects its Lc to that connection. Otherwise, it terminates the link and reports DL_RELEASE-ind to the NWK layer.

If a MAC_CON-ind primitive arrives in time but does not indicate a connection handover, the fixed part must assume that the connection is new, report to the NWK layer with a DL_RELEASE-ind that the old link has been dropped, and then report a new link with DL_ESTABLISH-ind.

At the portable, an involuntary handover is initiated when the DLC layer unexpectedly receives MAC_DIS-ind indicating abnormal release. As with the fixed part, the portable DLC layer has 10 seconds during which to issue a MAC_CON-req, wait for the MAC_CON-cfm and reconnect the current link to the new connection. If that fails, it reports DL_RELEASE-ind to the NWK layer and drops the link.

6.5.3 Connection Release

At the DLC layer, releasing a connection can be either expected or unexpected, resulting in the release of a connection being called *normal* or *abnormal*.

A normal release occurs in response to a DL_RELEASE-req primitive from the NWK layer to the DLC layer, containing the indication that the release is normal. For example, this happens on a normal call clear-down initiated by the user. In that case, the DLC layer (specifically, LAPC and Lc) attempts to send out any waiting control messages, control message segments in I-frames, or fragments of control messages, before actually initiating connection release. If, as a part of the normal release procedure, the buffered information cannot be transferred for any reason, the normal release can become an abnormal release.

An abnormal release occurs as a result of something going wrong, perhaps with the DLC layer's protocols or somewhere in the MAC layer, whenever the connection can no longer be sustained. If no data remain to be transmitted or the connection release is indicated as being abnormal, LAPC will initiate an immediate connection release with a MAC_DIS-req sent down to the MAC layer and indicate that to the NWK layer with a DL_RELEASE-cfm primitive.

LAPC then will terminate its operation and that of its partner Lc. The primitives used all include data to confirm to or inform the NWK and MAC layers whether the release was normal or abnormal.

At the peer DLC layer, a connection release is initiated from the MAC layer rather than the NWK layer. That comes up to the DLC layer as a MAC_DIS-ind primitive, again indicating normal or abnormal release. A normal release indication arises as a result of a normal release procedure at the other end, and the peer DLC layer terminates as described above, finally reporting that to the NWK layer. If the MAC_DIS-ind indicates abnormal release, that is, one not ordered normally by the peer DLC layer, an attempt at connection handover process will be initiated (see Section 6.5.2). Only if that fails will the link be dropped.

6.5.4 Broadcast and Paging

The broadcast/paging operation at the DLC layer is simple, because no upper-DLC entity is assigned for broadcasting. The DLC layer acts merely to pass on the broadcast messages, either upward or downward, in the order in which they are received. There is a means for the NWK layer to request "expedited" operation, in which a paging message that arrives is moved to the front of any queue that exists, but it is not required in GAP equipment.

Broadcasting a paging message (Figure 6.8) begins with the fixed-part NWK layer requesting it with a DL_BROADCAST-req primitive containing the data to be broadcast. The primitive also contains a list of one or more of the system's clusters across which the broadcast should be made. In expedited

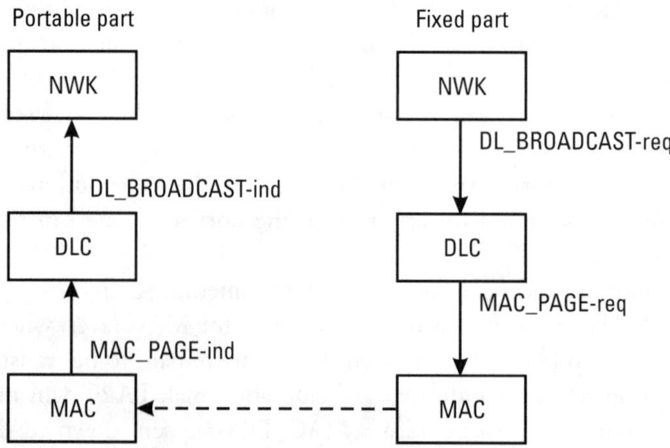

Figure 6.8 The flow of primitives in a broadcast (paging) operation. *Source:* ETSI [2].

operation, the paging/broadcast data arrives in a DL_EXPEDITED-req primitive instead. Whichever primitive is used, the DLC layer has to send the message and the paging identity separately to each cluster in the list, using several MAC_PAGE-req primitives carrying the same identities.

Whenever a portable's DLC layer, specifically the Lb, receives a MAC_PAGE-ind primitive containing the paging data, it immediately passes the data up to the NWK layer in a DL_BROADCAST-ind (or DL_EXPEDITED-ind) primitive, even if there was an error in receiving the message.

References

1. ETSI, *Digital Enhanced Cordless Telecommunications (DECT); Common Interface (CI); Part 4: Data Link Control (DLC layer) layer,* ETS 300 175-4, Sophia Antipolis, France, September 1996.

2. ETSI, *Digital Enhanced Cordless Telecommunications (DECT); Generic Access Profile (GAP),* EN 300 444, Sophia Antipolis, France, August 1997.

3. TIA/EIA, *Personal Wireless Telecommunications Interoperability Standard (PWT); Part 9: Customer Premises Access Profile (CPAP),* TIA/EIA 662-9-1997.

4. ITU-T Recommendation Q.920, *Digital Subscriber Signalling System No. 1(DSS1): ISDN User-Network Interface Data Link Layer—General Aspects—Digital Subscriber Signalling System No. 1,* Geneva, 1993.

5. ITU-T Recommendation Q.921, *ISDN User-Network Interface: Data Link Layer Specification—Digital Subscriber Signalling System No. 1,* Geneva 1993.

6. ETSI, *European Digital Cellular Telecommunications System (Phase 1); Mobile Station–Base Station System (MS-BSS) Interface Data Link Layer Specification (GSM 04.06),* TS 04.06, Sophia Antipolis, France, January 1995.

7

The Network Layer

7.1 Introduction

Once we have ascended the protocol stack to its top layer, the NWK layer, we have left behind the realm of frames, octets, bits, packets, radio carriers, and so on. There is a lot of specialized jargon in the lower layers of the protocol stack, invented as a shorthand for the mechanisms and concepts built on its foundation, the PHL. However, we are now in the realm of NWK-layer messages, which usually are recognizable to anyone with a little experience in mobile telecommunication services. That is because the NWK layer is really an information broker on behalf of the application it serves. Its main job is to get information from the application, assemble it into suitable messages, and get those messages to the other side of the link for the other half of the application to use.

Nevertheless, you will not escape protocol-specific jargon completely. To organize a complex set of operations and functions to cope with mobility support and call processing, the NWK layer introduces its own set. However, it is more intuitive, because the jargon relates to services with which you may be familiar, either provided by or supported by the protocol stack.

The NWK layer and its protocols (Figure 7.1) fit between the IWU and the DLC layer. The IWU is the formal adapter between the application and the top of the protocol stack. The application might be a portable handset with its display, keypad, microphone, ringer, and earpiece. Alternatively, it might extend as far as a large corporate PABX.

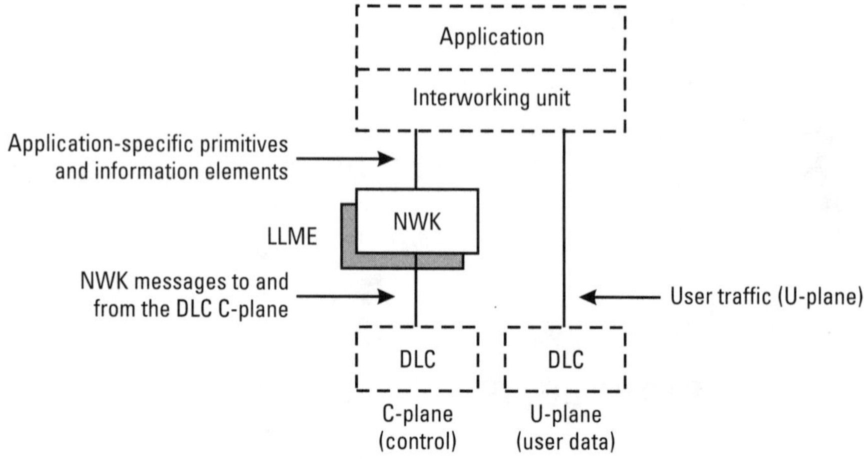

Figure 7.1 The NWK layer relative to the other protocol layers.

To do its job of serving the application, the NWK layer has to present a standard interface. Of course, the interface has to be general enough to support many different applications, so it cannot be specific to any one. That is precisely why there has to be an IWU between the application and the top of the protocol stack.

The main information structures the NWK layer deals with internally are messages, usually written as {MESSAGE-NAME}. A message is the primary information structure that is exchanged between the NWK layers in portable and fixed terminations. A message is defined by the NWK layer as containing a variable number of information elements (IEs), usually written as <IE-NAME>. Each IE contains a specific piece of information to be transferred. Most of the IEs are obtained from the IWU, which packages information obtained from the application into this standard form for the NWK layer to use. The IWU also unpacks information from the IEs it gets from the NWK layer and uses that information to drive the application. The NWK layer, therefore, is built around the definitions of those two information structures, the rules about how to construct and decode them, their uses, and their meanings.

Like the other parts of the DECT protocol, communication between the layers and communication with the LLME conceptually are carried out using primitives. Primitives transport messages and associate information through defined SAPs. The primitives and the SAPs are there to give structure to the protocol definitions and do not have to be implemented in practice. Primitives that are exchanged with the IWU contain groups of IEs from or for an application. Those that are exchanged with the DLC layer contain messages that have been assembled from those IEs according to NWK-layer rules.

As with the DLC-layer and MAC-layer standards, the NWK-layer standard [1] is a menu of features. You do not need them all for any specific service. For implementing voice service, you need only those features defined in the GAP standard [2]. In fact, you probably will get a better idea of the fundamental operation of the NWK layer by looking into the "NWK layer procedures" clause of the GAP rather than the equivalent parts of the NWK-layer standard. That is because the GAP procedures are not cluttered with full and general details. It is worth going through the GAP and the NWK standards with highlighter pens to match those parts referenced from the GAP in one color and the rest in another color. That shows just which parts of the NWK standard are needed in a simple voice system and those that are not.

7.2 Basic Principles

The first thing to note is that the NWK layer does not manipulate U-plane data directly. As can be seen from Figure 7.1, the NWK-layer U-plane is empty. Technically, user data, such as speech, passes without any intervention through the NWK layer, between a service access point connected to the IWU and one connected to the DLC layer. In real implementations, U-plane data may well flow directly between the DLC layer below and the IWU above, avoiding the NWK layer entirely, which is what Figure 7.1 shows.

That leaves the NWK layer in charge of only control information, with only C-plane protocol entities. Hence, only the C-plane is shown in the network layer reference model in Figure 7.2. The C-plane entities do, however, control when (or whether) the application is told that the U-plane is available (e.g., to allow it to mute the speech channel when there is nothing to pass on to the user).

The next thing to say about the DECT NWK layer is that it is in itself layered. The most important entities of the NWK layer are the services that the NWK layer provides via its higher entities: call control (CC), call-independent supplementary services (CISS), connection-oriented message service (COMS), connectionless message service (CLMS), and mobility management (MM). An application wanting to set up a call will communicate with the CC entity, exchanging IEs from or for the messages that CC exchanges with its peer at the other end of the link. If the application needs mobility support, it will talk to the MM entity.

All the higher entities are built on top of the link control entity (LCE), whose main job is to ask the DLC to set up a reliable C-plane link and then to

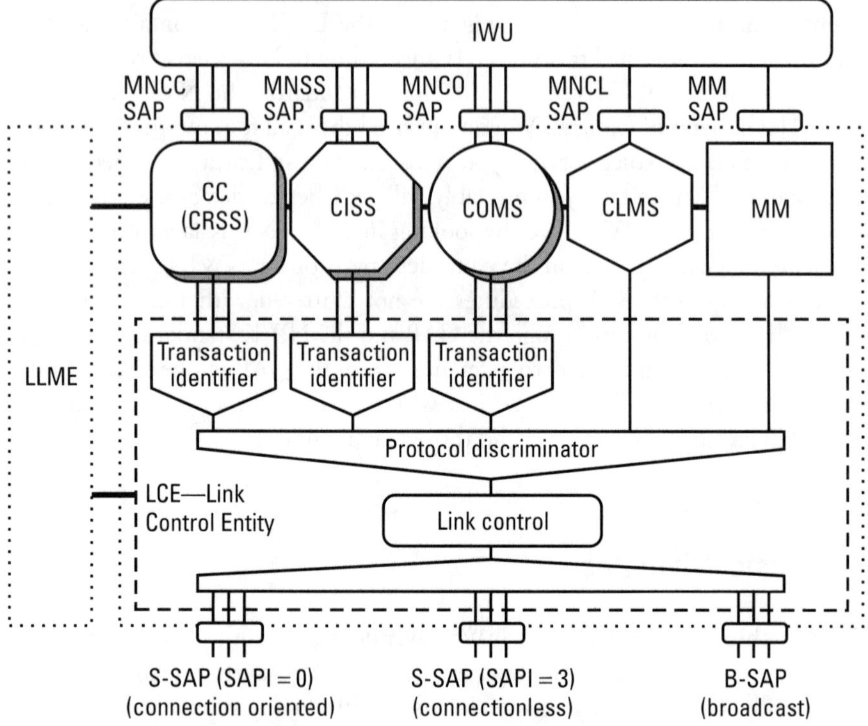

Figure 7.2 The NWK-layer reference model. *Source:* ETSI [1].

organize the flow of NWK-layer messages to and from the various higher entities. All of this structure is shown in the reference model in Figure 7.2.

All the higher entities and the LCE are introduced next, and some are explained in more detail later in this chapter. The detailed descriptions concentrate only on the entities needed for speech service using the GAP standard. They are:

- The LCE;
- The CC entity;
- The MM entity.

A simple GAP speech system does not need to implement any other higher entities.

7.2.1 Basic NWK-Layer Operation

The main job of the NWK layer is to organize information exchange on behalf of an application using the radio link. In a simple portable-originated telephone call, the handset has to tell the base station to go off-hook and what telephone number to dial. The NWK layer does that of course, but it also is designed to cope with much more complex situations on behalf of many other types of application. To do that, all the protocol entities in the NWK layer act by assembling NWK-layer messages from IEs that come from an application. The messages are sent to the peer NWK layer, where the IEs are unpacked and sent to the application at that end. In that way, the entities exchange essential information between the fixed-part and portable-part applications.

For each new call that it handles, the NWK layer creates an independent CC entity. CC is a message-driven call-control-state machine. It is a state machine because although there is a known set of IEs that have to be exchanged to set up a call, you do not know in advance in which order any particular application will supply them. You have to cope with all possibilities.

Although there can be many CC entities, there is always just one LCE. For each new CC, LCE orders the DLC layer to create a new radio link for the call, then it connects the CC entity to the new DLC link. Having done that, the LCE routes received messages upward from the DLC link to its corresponding CC. The LCE decides where to send messages based on information fields in each message called the protocol discriminator and the transaction identity. It also routes transmitted messages downward to the correct DLC link, based on the link's unique data link endpoint identifier (DLEI).

The NWK layer normally assumes that the link will be managed entirely by the DLC and lower protocol layers (including the management of bearer and connection handover). There is one exception: the case of external handover. In that case, handover optionally may be allowed between different DECT or PWT systems connected to the same network, and the NWK layer provides a mechanism to handle it.

A single MM entity provides mobility-management procedures to help an application track the location of portables, thus allowing calls to be delivered anywhere in a cellular coverage area. It also provides subscription services and mutual secure identification of fixed and portable parts and their users. It also provides encryption if it is implemented. The MM entity itself does not provide the database of portable locations, it simply provides the mechanisms to assist an application that needs to keep track of portable locations.

As a whole, the NWK layer provides an application with a set of services through a general-purpose interface into the protocol stack. The IWU has the job of translating that interface into something a specific application can recognize and work with.

7.2.2 The Link Control Entity

At the lowest level of the NWK layer is the LCE. A single LCE is responsible for controlling all C-plane DLC links and for supporting all calls active in the NWK layer. Its primary job is to manage the flow of information, in the form of network-layer messages, to and from the higher entities (CC, CISS, COMS, CLMS, or MM). In the case of a GAP system, most messages are from or for the CC and MM entities, although the LCE itself does handle a few messages.

The LCE is responsible for the first phase of call setup. When a higher entity is asked by an application to set up a call, it first of all asks LCE to obtain a reliable C-plane link from the DLC layer. After that, control of the link is handed over to the higher entity that requested it, and the LCE just routes the messages that set up, conduct, and close down the call. After setup, the NWK layer leaves the DLC layer to manage the link, except for the cases of (1) unexpected failure, when the LCE has to notify the higher entity that was using the link, and (2) optional external handover, when the CC entity has to reconnect a call via another system connected to the same network.

LCE has to handle two types of message labeled S-FORMAT and B-FORMAT. S-FORMAT messages enter and leave the NWK layer via the DLC interface called the S-Servive Access Point, or S-SAP, which is connected to the DLC layer's LAPC entity. They are normal control messages that contain explicit information about where in the NWK layer they come from or are to be sent. LAPC is able to protect those messages, ensuring they are neither lost nor corrupted by the radio link.

B-FORMAT messages enter or leave the NWK layer via the DLC interface called the B-Service Access Point, or B-SAP, which connects to the DLC's broadcast entity, Lb. They contain implicit routing information. Some of them, including {LCE-REQUEST-PAGE} for paging a portable when an incoming call arrives, actually are originated by or destined for the LCE itself.

For outbound messages, the LCE may have to set up several DLC links; thus the LCE has to select the correct one to which to send the messages. However, each new DLC link is given a unique DLEI for that purpose. If messages cannot be sent immediately, the responsibility for queuing them lies with the LCE.

7.2.3 Call Control

CC is the main protocol entity in the NWK layer. Each call has its own instance of CC distinguished from any others by its own transaction identifier value, allocated when the call is set up. CC provides a circuit-switched service that controls the U-plane services which provide user communications such as digital voice. The CC entity handles call establishment, service negotiation, call connection, call-information handling, service changes during an active call, and call release.

In setting up a call, CC first arranges for the LCE to set up a C-plane DLC link, then it uses that link to exchange messages with its peer in the opposite NWK layer, finally informing the IWU that the U-plane link can be connected to the applications at each end. Call setup involves the exchange of information so that the two ends of the link can identify themselves, verify that the access is allowed, and then set up the end-to-end link. That involves, among other things, the exchange of a portable identity, a fixed identity, and the called party number. Furthermore, the setup of a call in a system like DECT or PWT, which provides many different connection types for many different types of equipment, also involves the exchange of call attributes identifying the type of call required and the negotiation of which system features are supported at the two ends.

All the information needed to set up a call need not be exchanged in the first pair of messages sent by each peer CC entity. Further messages may be required. For that reason, a fairly complex state machine is defined by CC to ensure that all the relevant information eventually is exchanged between the two CC entities and is verified before the call can be connected.

Finally, of course, CC orders the LCE entity to release its link when the call is complete. That also may happen if any message exchanges fail to be completed within a specified time period, or if the information exchanged is not satisfactory.

7.2.4 Mobility Management

The management of mobility in DECT makes the basic assumption that a portable knows where it is, and informs the fixed part. The fixed part is not directly responsible for actively tracking all the portables whose incoming calls it handles, a principle that is more scalable than having the fixed part actively responsible for all the location tracking. A minor disadvantage is that different portables, perhaps from different manufacturers, are likely to have different performance, depending more on their design quality rather than the system's. Many people might expect the fixed system to be more in charge, but the principle is to devolve to the portable much of the management of mobility at all levels.

Having the portable manage its own mobility is achieved by the portable identifying from the fixed part's beacon the particular location area it is in. Once it has determined that it is in a new area, a portable tells the system where it is and then usually receives a temporary identity to use while it is in that area. A system's location area level identifiers are set up by the system installer and identify distinguishable areas that may be one cell or many, depending on how the system has been planned. A single-cell residential set clearly has only one location area. A small PABX with only a few cells also may have only one location area covering them all. However, a large-site PABX covering perhaps a few thousand people normally divides its many cells into smaller location areas, to spread out the traffic generated by incoming calls.

You may remember from Chapter 5 that the fixed part distributes paging messages indicating an incoming call for a portable over all the cells in at least one MAC-layer cluster. Hence, it is not necessary for a fixed part to know where a portable is to any finer precision than one cluster when delivering an incoming call. A location area is normally at least one cluster of cells, but it could be many clusters of cells, at the discretion of the system designer or installer.

The procedures provided by MM are designed primarily to manipulate temporary databases of handset location information, VLRs. However, more permanent databases, HLRs, are needed to support the management of identity and subscription. For supporting those registers, the MM entity also has procedures for identifying portables, including secure authentication of both the equipment and the user. It also provides means by which to register the portable as a user of a system by assigning it an identity under which it will be known to the system and giving the portable an identifier by which it can identify and lock to the system.

7.2.5 Supplementary Services

There are two distinct groups of supplementary services. The first group contains services that relate to a single call, the call-related supplementary services (CRSS). An example is explicit call transfer, which applies to the current call and requests the cordless system to transfer the call to another user. The second group, CISS, contains services that are independent of any specific call and relate to all (future) calls. An example of a CISS is "call forward on busy," which applies to all future incoming calls.

Because CRSS requests are call-related, the CC entity handling the call also handles CRSS requests during the call to which they relate. Therefore, there is no separate CRSS entity. A CISS request, however, requires that a special call be set up to invoke the service, so there is a CISS entity. While a normal call is set up by CC when the application orders the IWU to send the NWK layer a MNCC_SETUP-req primitive, the CISS call is set up by the application with an MNSS_SETUP-req, which causes a pair of CISS entities to be created instead. The CISS entities then handle the negotiation of the supplementary service—and nothing else—before releasing the radio link. Whichever of the two entities is set up, the NWK-layer messages used to invoke the supplementary service are the same.

There are three different ways of invoking a supplementary service. In the first procedure, the keypad protocol, a sequence of keystrokes is used to indicate the service required, and the portable's display can be used to indicate any response. An example might be the keying of *101234 during a call to transfer the call to extension 1234. The keystrokes simply are sent inside <SINGLE-KEYPAD> or <MULTI-KEYPAD> IEs within an appropriate message. The keystrokes are interpreted by the cordless application, and any response from the application may be sent back to the portable inside <SINGLE-DISPLAY> or <MULTI-DISPLAY> IEs. If it is a CRSS that is being requested, the IEs may be sent in a {CC-INFO} message during an established call. If it is a CISS request, the IEs might be sent inside the {CISS-REGISTER}, which requests the setup of a CISS call.

The second method is the feature key management protocol, which identifies certain feature keys on the handset with certain supplementary services. Pressing a feature key results in a <FEATURE-ACTIVATE> being sent to the fixed system coded with one of a specific list of features. The reply (if any) may be returned within a <FEATURE-INDICATE> IE, indicating what has happened as a result of the request.

The third method is the functional protocol. In this case, *functional* means that the portable-part and the fixed-part protocol stacks must have knowledge of and support for the supplementary service. There may be a special message for the service, such as {HOLD} or {RETRIEVE}. More generally, a <FACILITY> IE may be used, indicating supplementary services as coded in ETS 300 196-1 [3], which is sent in a {CISS-REGISTER} message.

7.2.6 Connection-Oriented Message Service

The COMS entity is not a part of a basic GAP speech product, so we will not describe it in detail. COMS provides a point-to-point packet service. It is used to provide a connection with more rapid call establishment than the CC entity (and it has a much simpler state machine than CC). COMS service includes the optional ability to rapidly suspend and resume the link, so releasing the lower-layer resources (e.g., the radio spectrum) during periods of inactivity. It is similar to a virtual packet-mode connection.

Like the CC entity, COMS is connection oriented and therefore has setup, information transfer, and release phases. The setup phase is used to negotiate service and to exchange portable and fixed identities, interworking attributes, and certain COMS C-plane attributes.

7.2.7 Connectionless Message Service

CLMS also is not required in basic GAP systems, so we will not describe it in detail either. It provides a connectionless point-to-point or point-to-multipoint single-packet communication service. Being a connectionless service, there is no setup or release phase to organize, and connectionless messages are data packets that are broadcast just on the basis that they may be received or not, depending on the radio propagation conditions. Each packet is a separate entity unconnected to any other as far as the CLMS entity is concerned.

There is only one CLMS entity, and it can receive or transmit a CLMS message at any time. As a connectionless service, the CLMS entity introduces no state machine of its own into the NWK layer. If the messages happen to be transmitted over a connection-oriented link, that will happen only if a link already exists; CLMS will not set one up.

There are two message formats, fixed and variable. Fixed format {CLMS-FIXED} messages are sent via the B-SAP to the DLC broadcast service (Lb). They only go from fixed to portable, using the B-FORMAT message structure. The message contains a fixed-format address section indicating the portable(s) for which the information in the data section of the message is destined. The variable-format messages use the S-FORMAT message sent via the S-SAP, which connects to the DLC's LAPC service. They can be sent in either direction. Again, the message contains a portable address, this time in a <PORTABLE-IDENTITY> IE. The rest of the {CLMS-FIXED} message can contain a sequence of IEs comprising the data to be transferred.

7.3 The Structure of NWK-Layer Messages

Messages are the highest level of information used in the NWK layer. They usually are identified by a name specific to the originating entity. For example, the {CC-SETUP} message originates from or is destined for a CC entity. Mostly, it is the higher entities that send and receive messages. The LCE entity sends and receives some messages itself (such as {LCE-REQUEST-PAGE}), although it usually just examines the headers of messages it receives and decides where it should route them.

As has already been said, there are two formats of message, S-FORMAT and B-FORMAT. S-FORMAT messages are exchanged with the DLC layer via the S-SAP (hence the name) and so use the DLC's LAPC protocol, which can provide error protection (see Figure 6.2). In GAP systems, those messages are used for normal call setup, information exchange, and call release. B-FORMAT messages use the B-SAP and hence pass through the DLC's broadcast service. They are used mostly to carry paging identities to deliver incoming call requests and are thus optimized for that use.

S-FORMAT messages have a more versatile structure than B-FORMAT messages. They are made up from IEs, such as <SINGLE-KEYPAD>. The reason for the separate definition is that most IEs are used by the IWU to allow an application to exchange information with the NWK layer, and they often are used in several different messages. The IEs permitted in a message and their required order are defined by the NWK-layer standard. The standard also defines which IEs appear in each of the primitives between the NWK layer and the IWU.

It should be remembered that most of the definitions of NWK layer messages have to do with their construction and transmission, not with their reception and interpretation. A wise implementer deals with transmitted messages strictly according to the standard but with incoming messages leniently if a sensible interpretation is available. Employing that principle prevents operational problems from arising in the future. It can save engineering rework costs if standards are updated and new equipment is released that takes advantage of new freedoms. It can even make your products look good by dealing properly with other poorly designed equipment.

7.3.1 Information Elements

There actually are three categories of IEs:

- DECT-specific IEs;
- DECT-standard IEs;
- DECT-transparent IEs.

DECT-specific IEs are those, such as <CONNECTION-ATTRIBUTES> that relate exclusively to the internal operation of the protocol and are not known about above the NWK layer. They never appear as a component of a primitive between the IWU and the NWK layer. The DECT-standard IEs are those such as <SINGLE-KEYPAD> that relate to the interaction of the protocol stack with the IWU and the application above it. DECT-transparent IEs, of which there are two (<IWU-TO-IWU> and <IWU-PACKET>) contain information relevant exclusively to the IWU. The protocol stack just transports that information transparently on behalf of the IWU. The last two categories appear inside primitives from IWU to NWK layer, as well as in messages from a NWK layer to its peer.

Regardless of category, there are two types of IEs: the fixed-length IE and the variable-length IE (Figure 7.3).

The fixed-length messages are identified by their having bit 8 of the first octet set to 1. The next three bits, numbers 5 to 7, identify whether the element is one of several 1-octet elements with a 4-bit information payload or a 2-octet element where the payload is in the second octet. The variable-length IEs are categorized by having bit 8 of their first octet set to 0. In that case, the second octet of the IE contains the length, *L*, (in octets) of the contents of the IE, that is, the length of the payload that follows.

A well-designed parser for NWK-layer messages is critical. First of all, it should be able to use general principles to walk through the contents of any

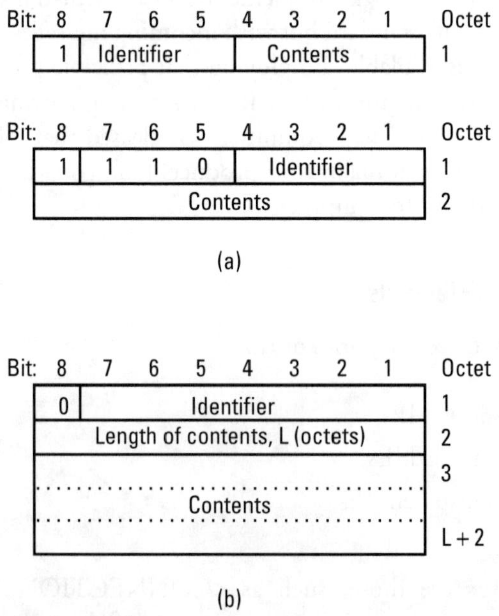

Figure 7.3 (a) Fixed-length and (b) variable-length IEs. *Source:* ETSI [1].

message containing a sequence of any fixed- and variable-length IEs in any order, independent of the message specification. Remember that in future, a new IE may be added to any message to implement a new feature. So decoding messages on the assumption that they always will contain just the IEs currently defined may lead to problems when interoperating with new equipment. The proper way to allow for that possibility is to be sure that you are able to parse a message into IEs independent of the actual IEs the message contains. Then you should simply ignore any IEs you do not recognize and interpret the rest.

The specific content of an individual IE has a defined structure that comprises a sequence of octets, each carrying a designated piece of information. A parser to determine the contents of an IE has to know the precise structure of that IE to decode it. However, a couple of general rules apply to decoding the contents of IEs. The first is that some "octets" may be extended to actually encompass a second octet when their highest bit, bit 8, is set to 0. That can happen more than once, but only for octets that the standard says can be extended that way. In that case, the so-called octets are numbered, for example, 5, 5a, and so on.

The other thing to note is that future extensions to IE contents may be possible by adding new octets to the end of an existing IE or by extending an existing octet that currently uses the extensible format. To be able to cope with that, a parser for any specific IE should be able to handle an IE whose as-received length is either greater than or less than the current specified value. In the case of an IE arriving that is longer than expected (perhaps from a newer piece of equipment), the parser should interpret only those octets about which it knows. In the case of an IE arriving that is shorter than the current standard shows (perhaps from an older piece of equipment), the parser should interpret only the actual octets received and should not simply expect to receive the extra octets it might send if it had constructed the IE.

Implementers should also beware of "spare" bits in IEs. If such bits truly are undefined, they can be assigned a meaning in future versions of the standard. Truly spare bits normally are set to 0 on transmit and must be ignored on receive. That is not the same as a field that is defined but has unassigned values. In that case, the field is set appropriately on transmit and must be fully decoded on receive. If an IE that is received contains a field value that an implementation knows nothing about, an intelligent decision must be made about how to proceed, depending on the individual case. The new field value indicates some information you know nothing of, so a decision must be made about whether associated data can be safely ignored. However, an implementation must not just fall over if field values are seen in IEs that were not defined when the system was designed.

In summary, it is important that an implementer take into account only the general principles when decoding a message into its IEs and when decoding an IE into its information contents.

7.3.2 S-FORMAT Messages

The S-FORMAT messages (Figure 7.4) are variable-length messages, made up from a header and a variable number of IEs. The precise contents of an S-FORMAT message may vary, but certain rules about its structure must be obeyed. Those rules include the ordering of both mandatory and optional IEs and which IEs are permitted in the message.

The general format of an S-FORMAT message is:

- Transaction identifier (TI) and protocol discriminator (PD);
- Message type;
- IEs.

The PD distinguishes which of the protocol entities in the NWK layer originated the message and which should receive it. That is the information the LCE uses to decide where to send a message it receives from the DLC layer.

The TI indicates which particular instance of the entity designated by the PD should receive the message. The LCE, the MM, and the CLMS entities are not allowed to exist multiple times, but the CC, COMS, and CISS entities may exist in many instances (Table 7.1). A separate CC entity, for example, exists for every active call at a fixed part (or portable part) that supports more than one simultaneous call. The transaction value in the TI-field identifies the correct entity to receive the message.

The TI+PD header occupies either 1 or 2 octets. One octet is enough for identifying transaction values up to 6, while the 2-octet version is needed to

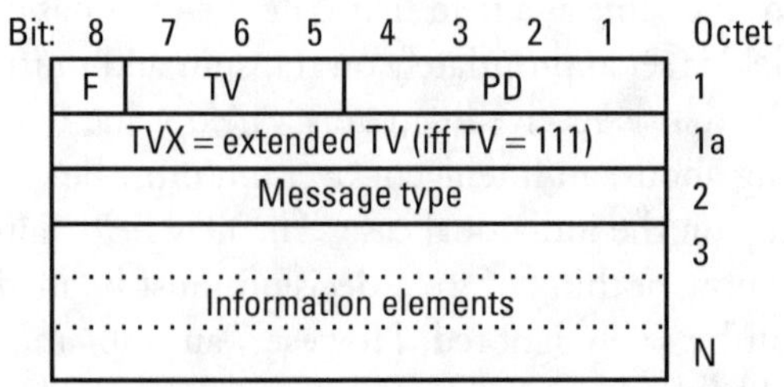

Figure 7.4 The S-FORMAT message. *Source:* ETSI [1].

Table 7.1
Allowable Transaction Values

Protocol Discriminator	Maximum Number of Transactions	Allowable Transaction Values	Comments
CC	263	0 to 6 then 0 to 255 extended	The extended transaction value is used if there are more than seven calls
CISS	7	0–6	Up to seven CISS entities
COMS	7	0–6	Up to seven COMS entities
CLMS	1	0 only	Just one CLMS entity
MM	1	0 only	Just one MM entity
LCE	1	0 only	Just one LCE

identify CC transactions in systems having seven or more different simultaneous transactions. Even if a system supports only up to six calls simultaneously, the handset may be asked to work with systems that support more. Hence, all equipment has to support the extended TI+PD format as well as the short form.

The coding rules also give the permitted IEs for any message and designate which are mandatory and which are optional. The mandatory IEs always come in the first section of the message, and the optional IEs come in the second section. Furthermore, the IEs in each section are always transmitted in ascending order of IE identifier. That allows a particular IE to be found or eliminated as not found without a system having to scan all the IEs in a message. Certain fixed-format IEs, such as <REPEAT-INDICATOR> may appear in front of any IE, so implementers should expect fixed-length IEs at any point in an S-FORMAT message.

It is important to realize that the restriction that specifies which IE is allowed in each message is a restriction on what may be sent and does not automatically imply any rule on whether you should or should not be capable of receiving messages containing other IEs. It has to be taken into account that the list of permitted IEs in any message may be amended in the future, as the DECT standard evolves to deal with more applications or to correct mistakes. It is essential to assume that will happen and that only the general rules about S-FORMAT message construction will remain invariant, not the specific contents of the messages.

7.3.3 B-FORMAT Messages

The B-FORMAT messages (Figure 7.5) are specified as simple fixed-length structures intended to convey information to or from the slow broadcast channel (B_S) in the MAC. This channel is mainly used for broadcasting paging messages to a portable indicating the presence of an incoming call, as shown in Figure 7.5(a). Hence the usual information content of a B-FORMAT message is a portable identity. This is commonly the short temporary portable user identity (the TPUI) which is assigned to a portable when it notifies the fixed system of its presence (see Section 7.6.1). However, a B-FORMAT message also may contain a longer identity and indeed may contain other broadcast information, for example, broadcast alphanumeric display data, as shown in Figure 7.5(b).

The normal fate of a B-FORMAT message from the NWK layer is transmission in a DL_BROADCAST-req primitive via the B-SAP, which sends the message to the DLC's broadcast entity, Lb.

7.4 Link Control Entity Procedures

The most important procedures of the LCE are those involved with the setup, maintenance, and release of DLC links on behalf of the NWK layer's higher entities and the routing of messages to and from their correct destinations.

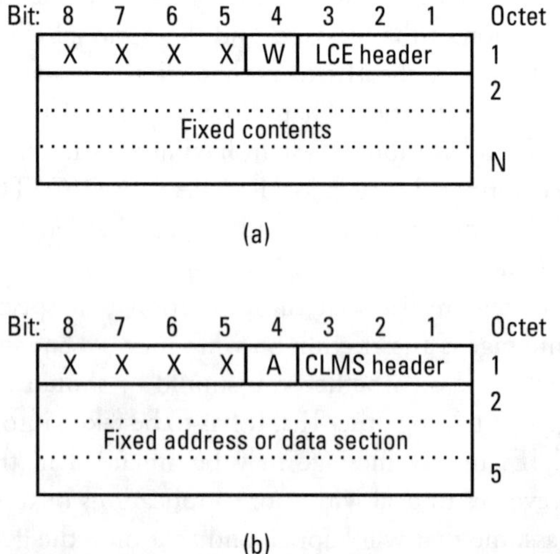

Figure 7.5 (a) B-FORMAT {LCE-REQUEST-PAGE} message; (b) B-FORMAT {CLMS-FIXED} message. *Source:* ETSI [1].

It is important to distinguish between calls and links. A link is a service the DLC provides to the NWK layer, and it is the NWK layer's LCE that is responsible for driving the DLC layer and setting up links for the rest of the NWK layer (the higher entities) to use. If appropriate, it is a higher entity (e.g., the CC entity) that sets up a call using a link. Some of the higher entities use a link but do not set up a call. The connectionless message service, for example does not set up calls. The distinction is also important because putting one call on hold, for example, allows the link to be used again for another call. When the second call is dropped, the link may be retained to reconnect the held call. In that case, only one link is needed for the two calls.

7.4.1 Link Setup From the Portable

The LCE's role in call setup is to create a DLC link of the type specified in the request it receives from a higher entity. It then allows the higher entity access to that link by routing its messages to and from the specific DLC link it created. It is the higher entity that completes the call setup using the new DLC link. In the case of GAP equipment, it is the CC entity that completes and then controls the call. In the descriptions that follow, CC will be assumed to be in charge. However, it could be another of the higher entities controlling the call if the call is for data exchange or some other type of nonvoice service.

The actions of the LCE in call setup from the handset and the events leading up to the LCE's actions are described next. Remember that for incoming calls to the portable, the same procedure is prefaced by the fixed part paging the portable with the {LCE-REQUEST-PAGE} message before the portable itself starts the link setup process described in Section 7.4.2.

Call setup starts when the cordless application requests a call to be set up via the IWU. The user of a portable handset may have pressed the off-hook button, for example, and expects to hear dial tone shortly afterward. The IWU translates that request into a primitive for one of the higher entities, asking for a certain type of link. For a speech call originating at a portable, the primitive will be a MNCC_SETUP-req, and the NWK layer's LLME will set up a new CC entity with a unique transaction identifier. The application itself does not know about the DLC layer's class A or class B links or about using half, single, or double slots at the PHL, but it is the job of the IWU to translate the simple "off-hook" command into the correct request to the NWK layer. In a simple telephone, the IWU has to know that the "off-hook" request from the particular application asks for a connection-oriented duplex link using the DLC layer's class A LAPC and a pair of single slots at the PHL (i.e., 32-kbps user data in each direction). Special codes in the <BASIC-SERVICE> IE supplied with the MNCC_SETUP-req primitive have been assigned as shorthand for several common types of link, including a normal telephone call. The CC entity puts the <BASIC-SERVICE> IE (and some others) into a {CC-SETUP} message, which is sent to the LCE to be forwarded to the CC entity at the base station.

On receipt of the message, the LCE takes action to set up a DLC link for the call. The LCE obtains a new, unique DLEI from the LLME, and it orders the DLC layer to set up a reliable C-plane link to the other DECT part. It sends the DLC a DL_ESTABLISH-req primitive that contains the requested link characteristics and the DLEI so that the new link can be identified in the future. The primitive is issued to the DLC layer through the connection-oriented S-SAP, and when the link is set up, the {CC-SETUP} message the primitive contains is handled by the DLC layer's LAPC entity (see Sections 6.4.1 and 6.5.1).

Issuing the DL_ESTABLISH-req to the DLC takes the DLC from the "link released" state into the "establish pending" state (Figure 7.6). At the LCE's option, the primitive may optionally not contain the {CC-SETUP} message, but the fixed part CC entity will expect it to be sent in a DL_DATA-req primitive soon after the "link established" state is reached.

The DLC now attempts to set up the type of link requested. For GAP speech, that will be a class A link. (Note that Figure 7.6 shows only those DLC states that are mandatory for GAP applications; suspension and resumption of a link are available in class B DLC links but are not required by GAP.) While the link establishment is pending, the LCE has to store in a message queue any further messages it receives, to be sent out when (or if) the link finally gets set up.

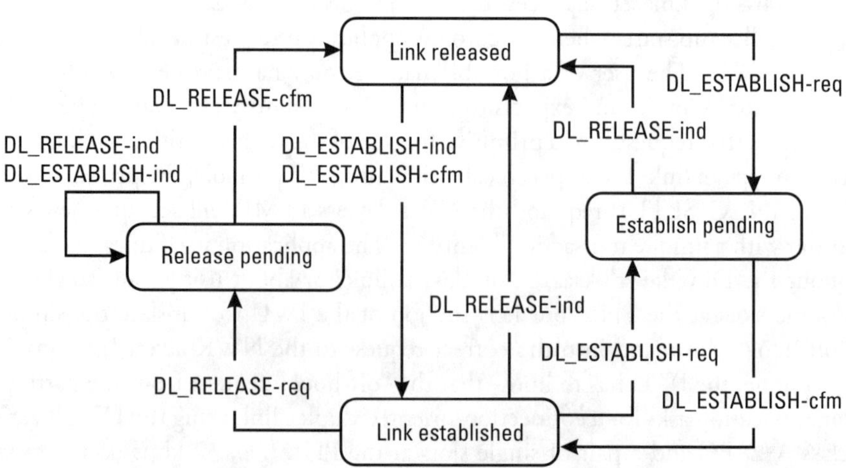

Figure 7.6 The states of the DLC as seen by the LCE during call setup. *After:* ETSI [1].

If a DL_ESTABLISH-cfm is received from the DLC layer, the DLC has moved to the "link established" state. At that point, the LCE has done its job in setting up the link and leaves the rest of the call setup for the two peer CC entities to sort out by exchanging messages (covered in Section 7.5). Those messages are sent by the LCE in DL_DATA-req primitives via the connection-oriented S-SAP for the DLC LAPC entity with the corresponding DLEI to transmit.

At the fixed-part LCE, the arrival of a DL_ESTABLISH-ind primitive from the DLC layer indicates that the fixed part has detected a portable's attempt to set up a link. However, the fixed-part LCE does not consider the link setup to be successful until the first NWK-layer message also has been received. That usually is a {CC-SETUP} message arriving with the first primitive; if it is not, the fixed-part LCE waits for a short time (designated <LCE.05> and having the value 5 s) for the message to arrive. If the message arrives, the fixed-part NWK-layer LLME starts up a new CC instance to complete the call setup in collaboration with its peer at the portable. If not, it orders the DLC layer to terminate the link by sending it a DL_RELEASE-req primitive containing an "abnormal release" reason.

7.4.2 Link Setup From the Fixed System

At the fixed part, a request to set up an incoming call initially is handled at the NWK layer in the same way as a request at a portable. However, the arrival from the network of a {CC-SETUP} message triggers the fixed part's LCE to transmit an {LCE-REQUEST-PAGE} message within a DL_BROADCAST-req primitive, sent to the DLC broadcast entity via the B-SAP. The LCE keeps hold of the {CC-SETUP} message until a recognized portable responds to the page. Portables that declare they support "FAST-PAGE" capability can be contacted more quickly by paging them with a DL_EXPEDITED-req primitive. However, that is not required for voice systems. It is useful for data systems but does not work with normal GAP portables.

For the fixed system to be able to page a portable, it previously must have attached to the fixed system using the appropriate MM procedure. As a result, the fixed part knows precisely to which section(s) of the system to route the paging request to reach the portable. Also, the MM attach procedure usually results in the portable being assigned a TPUI, which will be the identity used to address the portable in the {LCE-REQUEST-PAGE} message.

After issuing {LCE-REQUEST-PAGE} to the DLC layer, the LCE waits for <LCE.03> = 3 s for a DL_ESTABLISH-ind primitive. The portable is required to respond to the paging request with a DL_ESTABLISH-req primitive sent to its own DLC layer, just as if the portable were originating the call.

However, in the case of a paged portable setting up a link, the primitive must contain an {LCE-PAGE-RESPONSE} message revealing the portable's full identity, so the fixed system can check that it really is the portable it wants to contact. The fixed system also has to check which of the possibly many outstanding {LCE-REQUEST-PAGE} messages it is to which this response is directed. If DL_ESTABLISH-ind is received in time by the fixed part, with an {LCE-PAGE-RESPONSE} message containing a recognized portable identity, the DLC link is now in the "link established" state. The fixed part's LCE sends the queued {CC-SETUP} message via the connection-oriented S-SAP using a DL_DATA-req primitive. As in the case of portable-originated call setup, the CC entities now complete the call setup process (covered in Section 7.5.2).

Note that for incoming calls, the link is fully set up by the LCE using this procedure before the user decides whether to answer the call. The established link is used for alerting the portable. Completing the call setup and connecting the user, however, happens only if the alerting signals are answered.

7.4.3 Link Maintenance, Suspend, and Resume

The maintenance of an active link is the responsibility of the DLC layer, not the NWK layer. Within a call, the role of the LCE is limited to routing messages from the controlling entity to the DLC, based on their DLEI, and routing messages from the DLC to the controlling entity based on their TI and PD fields.

The LCE may, however, receive an unexpected DL_RELEASE-ind primitive at any time if the link is broken for any abnormal reason. It is the LCE's job to pass that information on to the relevant higher entities for them to decide whether to do anything. The LCE does not attempt to reestablish the link on its own. The mechanism by which the LCE notifies the higher entities is an internal NWK layer matter and is not specified.

The DLC's link-suspend and link-resume procedures are not required by GAP systems. They are useful for releasing and reacquiring lower-layer resources, especially the radio spectrum, in services where data transport is intermittent, and releasing the spectrum when it is not needed is the friendly thing to do. That does not apply to speech calls.

7.4.4 Link Release

A link is released on request of a higher layer entity or whenever a call-setup timer has expired before the call setup is complete. A normal call release is initiated in a GAP system by the CC receiving a MNCC_RELEASE-req primitive from the IWU, perhaps in response to the portable user pressing the "on-hook" button. It then requests the LCE to release the underlying link.

LCE is notified that one of its links no longer is needed via an internal NWK-layer communication referred to in the standard as NLR ("no link required"). It is not specified, because it is internal to the NWK layer, except for the fact that it must indicate to the LCE if the link should be released immediately or should be retained for a while for some possible reuse within a short period of time. In the latter case, the NLR indicates "partial release," which is used by the CC service, for example, to indicate that a held call exists and may be reconnected shortly via the existing link.

If LCE receives NLR without the "partial release" indication and no other higher entities currently are using the link, the link is released immediately. That may occur, for example, if a portable user hangs up a single active call. The IWU translates this application-specific action into a MNCC_RELEASE-req primitive to the CC entity, and CC then indicates NLR (without "partial release") to the LCE.

The LCE then sends a DL_RELEASE-req to the DLC containing a "normal" release parameter. The link state then moves to the "release pending" state, and the LCE waits for up to <LCE.01> = 5 s for the DLC to respond with DL_RELEASE-cfm, which usually confirms "normal" release if the release process worked correctly. If <LCE.01> expires before the primitive arrives, the LCE sends another DL_RELEASE-req with an "abnormal" release reason, which the DLC is required to act on immediately.

If "partial release" was indicated in the NLR notification, it is expected that some other higher entity will be using the link within a short period. The LCE waits for <LCE.02> = 10 s and after that keeps the link running only if any other higher entity has started using it. If the link is not in use after all, it is released using the "abnormal" reason.

7.5 Call Control Procedures

Generically, the CC procedures support the setup, maintenance, and release of both circuit-oriented and packet-oriented services. GAP systems, however, use only circuit-oriented calls using basic connections. Packet-mode procedures are available if the MAC and DLC layers support advanced connections, but they are not used to carry speech.

Each call that is set up is controlled by a separate instance of the CC entity. Hence, if multiple calls are supported, there can be multiple instances of CC at the fixed part. That can be true even at a portable where you may have one call on hold while you deal with another. Each CC is addressed in messages not only by a PD indicating that a message is for a CC but by a TI indicating to which specific CC instance it should be sent.

Multiple active calls need multiple DLC links that are identified via a unique DLEI that is assigned to each at call setup and used by the CC entity and the LCE to get NWK-layer messages sent via the right link. In simple instances, however, especially simple portable terminals, it may be that only one CC instance is ever operating, when only one call at a time is supported.

The CC entity handles CRSS as well as basic call control. It also has procedures for changing attributes in a connection, such as the bandwidth used, service rerouting, service suspension, and service resumption. The procedures for making such changes require that an advanced connection be established by the DLC and MAC layers. That is not a requirement for GAP-compliant systems, so the procedures are not covered here.

7.5.1 The Call Control State Machine

The primary function of the CC entity's call-establishment procedure is to get the relevant termination, fixed or portable, from what is called the NULL state to the ACTIVE state. Since the CC state machine is driven by NWK-layer messages, a prerequisite for the functioning of the CC entity is that the DLC layer has established a reliable C-plane link. It is the responsibility of the LCE to first set up that link before a CC entity can use it to set up a call (see Section 7.4.2).

The NWK-layer standard [1] provides both state transition tables and state transition diagrams for the fixed and portable parts, indicating which messages are required to make the transition via some or all of the intermediate states. Figure 7.7 and Figure 7.8 are cut-down versions of the state transition diagrams for the fixed and portable parts, respectively, indicating the most protracted sequence of events that can get enough information exchanged between the fixed and portable parts and the called party to get a call set up that originated at the portable.

The state diagrams make it easy in principle to design a CC entity. The biggest headache an implementer is likely to have is that the contents of each message have to be matched to the specific requirements. In that respect, the GAP standard [2] contains tables that specify the required contents for voice systems.

A key thing to note about handling the CC state machine is that you have to decide in each state what to do with the unexpected event as well as the expected and specified. Handling of errors and exceptional cases is specified for each entity. In the case of the CC state machine, all unexpected messages except {CC-RELEASE} and {CC-RELEASE-COM} are ignored, so protocol errors

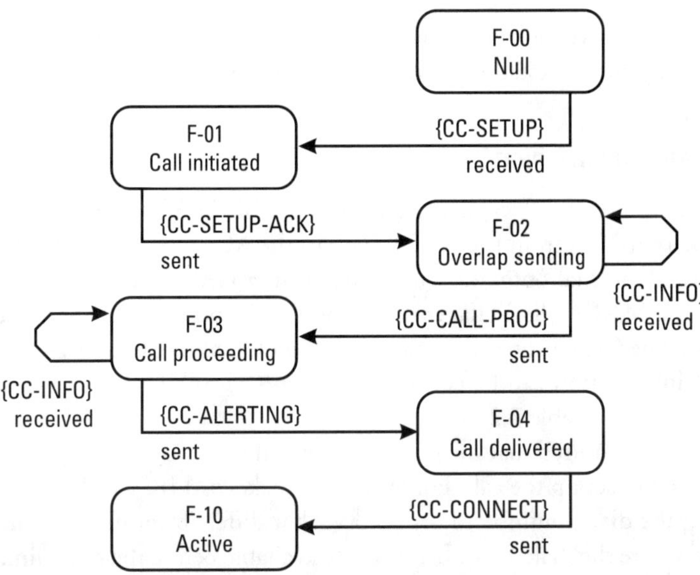

Figure 7.7 Call setup state transitions at the fixed part.

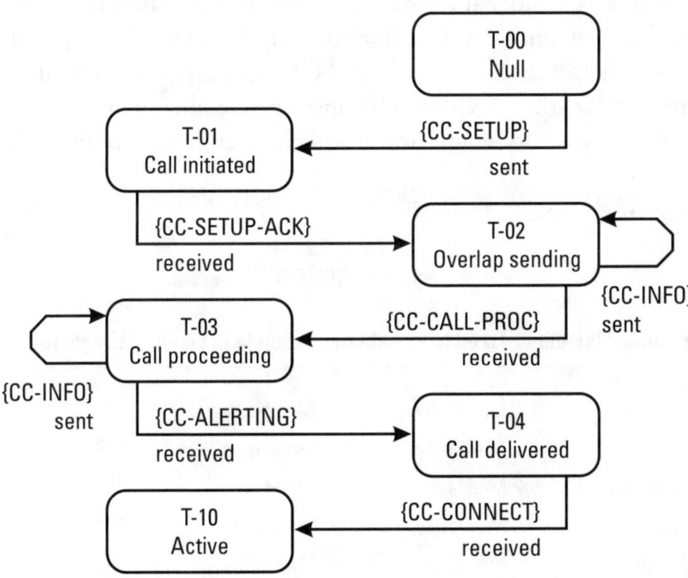

Figure 7.8 Call setup state transitions at the portable part.

are handled by the built-in timeout requirements. In the case of those two messages, they initiate a normal or an abnormal release procedure, respectively.

An alternative look at the paths through the CC state machine is taken in Section 7.5.2 for call establishment and call release.

7.5.2 Call Establishment

The GAP standard [2] explicitly shows some of the message sequences that cause the transition from the NULL state to the ACTIVE state in the CC state machines. A normal outgoing call set up from a GAP portable is initiated on receipt of a MNCC_SETUP-req from the IWU, specifying that a basic call is required. The CC entity passes that on to the LCE as a {CC-SETUP} message. A lot of information can be contained in the {CC-SETUP} message, some of which is listed in Table 7.2.

It is not required that the portable signal in the {CC-SETUP} all information needed to set up the call. For example, the keypad IE, which is one way of signaling the dialed number, is optional. That differs from most cellular radio systems, where the dialed number must be available before the radio link can be set up. The difference here is that in DECT and PWT the user is allowed to hear dial tone over the air if required. That costs a little in spectrum usage, of course, but it is a penalty paid to keep the operation identical to that of a wired telephone. The remaining information to set up the call, perhaps just the dialed number, is sent in subsequent {CC-INFO} messages. However, if an implementer required it, that {CC-SETUP} message could contain all the information a fixed part needs to set up the call and connect the user to the called party.

Table 7.2
Some Elements of the {CC-SETUP} Message

Information Element	Fixed to Portable	Portable to Fixed	Length (octets)
PD	Mandatory	Mandatory	1/2
TI	Mandatory	Mandatory	1/2
Message type	Mandatory	Mandatory	1
Portable identity	Mandatory	Mandatory	5–20
Fixed identity	Mandatory	Mandatory	5–20
Basic service	Mandatory	Mandatory	2
Keypad	Not allowed	Optional	2 or more
Feature activate	Not allowed	Optional	3–4

On receipt of a {CC-SETUP} message at the fixed-part CC entity, an MNCC_SETUP-ind primitive is sent to the IWU. Several replies from the fixed-part application layer are possible:

- MNCC_SETUP_ACK-req is the response if there is not enough information in the {CC-SETUP} for the fixed system's local network to connect the call. The fixed-part CC entity responds by sending a {CC-SETUP-ACK} message to the portable. It now expects the rest of the call setup information (typically just the called-party number) to be sent in one or more {CC-INFO} messages before it can proceed. There may be dial tone connected at this point.
- MNCC_CALL_PROC-req normally is indicated only when there is enough information from the portable for the fixed part to commence connecting the call. It may be indicated if some of the called-party number is still missing, so remaining numbers could be dialed and sent in <KEYPAD> IEs within {CC-INFO} messages. In response to this primitive, the fixed-part CC entity sends a {CC-CALL-PROC} message to the portable. In common with fixed telephones when dialing has started but is not complete, dial tone will have been removed at this point.
- MNCC_ALERT-req is the response if all information has been supplied in the {CC-SETUP} (and any subsequent {CC-INFO} messages) and the called terminal has been contacted and is believed to be ringing. The fixed-part CC entity replies to this indication with a {CC-ALERTING} message. There may be ringing tone available at this point.
- MNCC_CONNECT-req is indicated if all information has been supplied in the {CC-SETUP} (and any subsequent {CC-INFO} messages) and the call has been set up to the destination and has been answered by the called party. The fixed-part CC entity sends a {CC-CONNECT} message in response.

Having issued the {CC-SETUP}, the most direct response possible from the fixed part is the {CC-CONNECT} message, which would take the portable directly to the ACTIVE state. More typically, the call setup procedure proceeds through some of the intermediate replies list above and may proceed through them all (Figure 7.9). The minimum information you have to provide and exchange by following that (or any other) path through the state transition

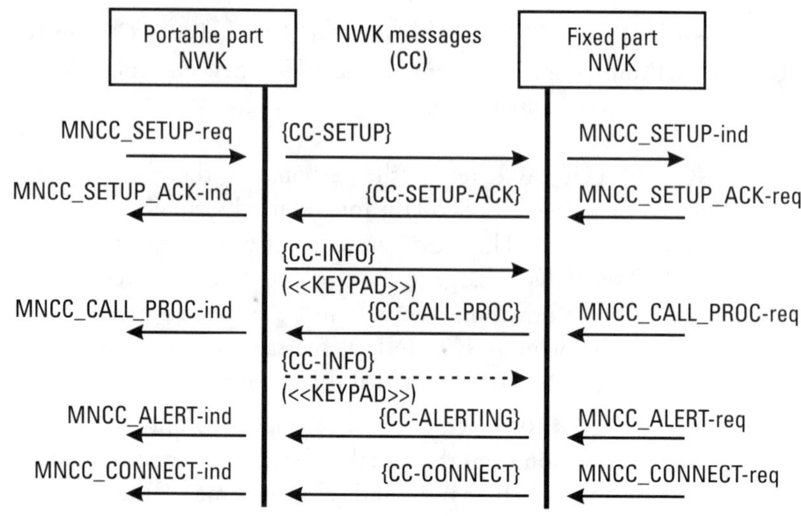

Figure 7.9 A maximum outgoing call-setup sequence. *Source:* ETSI [2].

table may be extracted from the mandatory IEs listed in the standards for each message.

Figure 7.9 corresponds with the call-setup sequences in Figures 7.7 and 7.8. By comparing the three figures, you can connect the sequence of primitives, the sequence of messages, and the sequence of states at both the fixed and portable parts.

Things also can go wrong in setting up a call. A radio system is not totally reliable, so you always have take that into account. Various timers are used to take a failed call setup back to the NULL state if something goes wrong. Furthermore, the portable's call-setup attempt might be rejected by the fixed part. That may happen if the portable asks for more resource than the fixed part has available, or if the portable asks for features that the fixed part cannot supply. The full state transition tables in the standards contain means for that to happen by the use of the {CC-RELEASE} or {CC-RELEASE-COM} messages.

For incoming calls to a portable, the most protracted incoming call using mandatory CC states is shown in Figure 7.10. The fixed part's LCE, as we saw in Section 7.4, deals with the preliminaries of paging the appropriate portable with the {LCE-REQUEST-PAGE} message. It holds the {CC-SETUP} message in its buffers and sends it only when a {LCE-PAGE-RESPONSE} is received that it fully recognizes, thus beginning the illustrated call setup sequence.

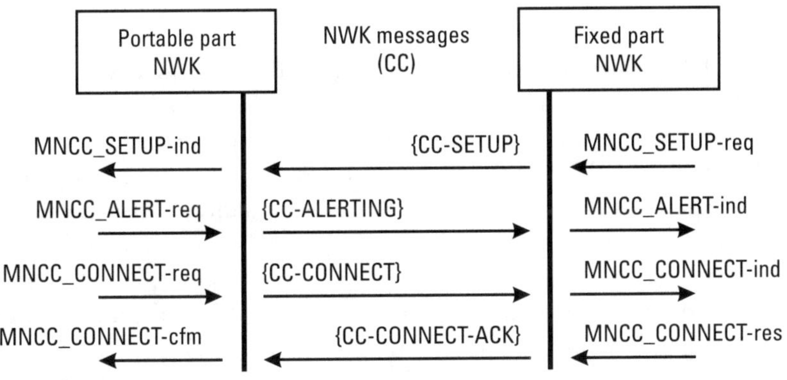

Figure 7.10 A maximum incoming call-setup sequence.

On receipt of the released {CC-SETUP} at the portable, the portable's application layer receives a MNCC_SETUP-ind from the CC entity and may respond to the primitive with one or both of the following:

- MNCC_ALERT-req, which usually indicates that alerting (ringing, vibrator, visual bell) has been turned on and the user may be aware of the incoming call. The portable-part CC entity then sends a {CC-ALERTING} message to the fixed system.
- MNCC_CONNECT-req, which usually indicates that the user has accepted the call by going "off-hook." The portable-part CC entity then sends a {CC-CONNECT} message to the fixed system.

It normally is expected that there will be enough information in the {CC-CONNECT} from the fixed part to dispense with any of the other intermediate responses that are possible in the case of an outgoing call. For that reason, the portable states associated with them in the full state diagram in the standards are not mandatory.

After the {CC-CONNECT} when the user does go off-hook, the final step in incoming-call setup is for the fixed part to send {CC-CONNECT-ACK}, after which the incoming-call setup is complete.

7.5.3 Call Release

An application that wants to release a call indicates that to the CC entity with a MNCC_RELEASE-req primitive sent to the NWK layer via the IWU. Call release is a simple procedure initiated by either CC entity sending a {CC-RELEASE} message and waiting for a {CC-RELEASE-COM} in reply. If the reply is not received within a period called <CC.02> whose value is 36s, the CC entity attempts to force its peer to release the call by sending a {CC-RELEASE-COM} and simply dropping back to the NULL state. Receipt of an unexpected {CC-RELEASE-COM} message indicates that the peer CC had previously tried but failed to clear the call and always causes the CC entity to terminate the link.

In a normal call termination, receipt of a {CC-RELEASE} message causes the receiving CC entity to ask the IWU with a MNCC_RELEASE-ind primitive if it may release the link. If the IWU accepts the release, it indicates that with MNCC_RELEASE-res, causing the CC entity to reply with {CC-RELEASE-COM} and enter the NULL state.

7.6 Mobility Management Procedures

The MM service in the NWK layer is what differentiates DECT and PWT from simple one-cell cordless telephones. It may seem too obvious to state, but fundamentally, the DECT model of mobility makes the assumption that only portable parts can be mobile. Therefore, it provides the means for fixed parts to track the locations of portables to be able to deliver incoming calls. However, it must be remembered that mobility is really an application-specific matter. The MM procedures are designed to support an application's mobility requirements rather than provide them all. Hence, you will not see here procedures for the management of HLRs or VLRs, just the procedures and data structures that allow the protocol stack to support an application that might want to implement such features.

To implement a simple single-cell residential telephone, you may reasonably think it is not necessary to implement much in the way of MM. After all, a single-cell system does not have any mobility outside its own cell. However, the MM procedures encompass a number of operations that can be used in support of MM but that still are essential for even a single-cell system. They include the authentication, identification, and access rights procedures, all of which are required to ensure that a user does not gain access to telecommunications services at someone else's expense. The MM entity also provides procedures for engaging and disengaging ciphering if that is supported.

One of the key things to understand is that the MM entity deals with two types of information. The first essentially is temporary information that arises out of the portable's movements. That information is dealt with through the location procedures and the identity procedures. In essence, those procedures manipulate data in the VLRs of a system that keep track of portable locations and temporary identities. The second type of information is more permanent information, which belongs in the fixed system's HLR and in nonvolatile storage in the portable. That comprises the access rights information and authentication keys, which even a residential cordless telephone set has to store.

7.6.1 Location Procedures

The basic reason for providing a suite of location management procedures is to allow incoming calls to be delivered to a portable. In the case of outgoing calls, the portable already knows where it is and just attempts to contact the nearest recognized cell. There are, in fact, two types of location requirement:

- *Location*, to inform the fixed system of the location of the portable in terms of the fixed system's assigned location areas;
- *Attaching and detaching*, to inform the fixed system that the portable is or is not ready to receive incoming calls.

The functions of those two types of procedure overlap. An attach message carries with it an implicit location function, since the portable informs the fixed system of its location just by attaching. There also is the use of a location procedure when the portable already has attached to the fixed system but has changed location. The notification of a change of location carries an implicit attach. As a result, procedures called simply *attach* and *detach* cover both of the above requirements.

A third procedure in the location procedures suite, called *location update*, informs the portable that the fixed system has updated its location area identifiers, inviting the portable to reattach and reconfirm its location.

7.6.1.1 Attach

Attach is used on two occasions. The first is the initial contact between a fixed system and a portable that has just been switched on or a portable that has just arrived within the service area of a system where it is permitted to receive service.

We saw in Chapter 5 that on power-up a portable searches for the best base station it can find that is broadcasting an ARI it recognizes. The portable recognizes the ARI broadcast from the fixed part by comparing a specified part of it to a portable access rights key (PARK). The PARK gets stored in the portable as part of the portable subscription process, which tells it which system(s) it may access. The immediate purpose of identifying a good base station that will give it service is for the portable to attach to the fixed part and by doing so inform the fixed part, first, that it is within the system at a particular location and, second, that it is ready to receive incoming calls.

The second function of attach is to inform a fixed system that the portable has moved to a new location area and expects to receive incoming calls there instead of at its old location. The same attach procedure is used for both purposes. The only difference is that different IEs are contained in the messages exchanged between the fixed and portable parts.

The initial attach procedure is initiated whenever the NWK layer at the portable receives an MM_LOCATE-req primitive from the portable application. In response to that, the portable's MM entity sends out over the air a {LOCATE-REQUEST} message containing a <PORTABLE-IDENTITY> IE declaring the user's basic unique identity, the international portable user identity (IPUI) (Figure 7.11).

If an attach procedure is intended to inform the fixed part that the portable's location has changed, then extra IEs are added to the {LOCATE-REQUEST} message to identify the previous location to which the portable was attached. That extra information includes a <FIXED-IDENTITY> IE containing the ARI from the previous location area and a <LOCATION-AREA> IE containing the old location area level identity (discussed in Section 8.4.3).

On receiving a {LOCATE-REQUEST} message for either an initial attach or for notification of a new location, the fixed part issues an MM_LOCATE-ind primitive to the IWU and hence to the mobility application. It then awaits confirmation or rejection of the location request. Receipt of an MM_LOCATE-res containing an "accept" indication allows the fixed part NWK layer to respond with a {LOCATE-ACCEPT}. That indication contains a <LOCATION-AREA> IE informing the portable of the specific location area inside which it need not reattach, plus a <PORTABLE-IDENTITY>

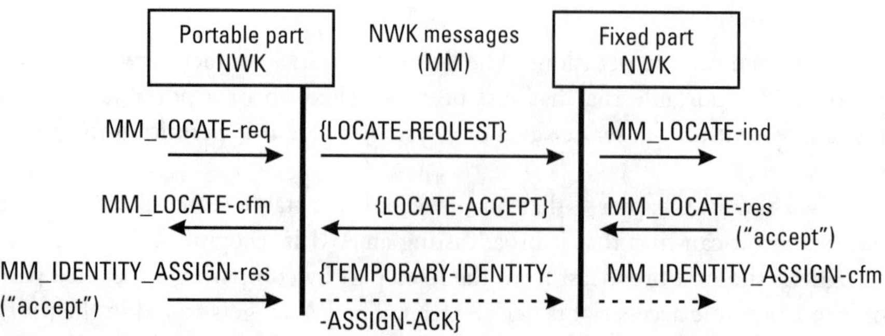

Figure 7.11 Attaching to a network. *Source:* ETSI [2].

which allocates the portable a TPUI, which is the shortened temporary identity that from now on will be used to page the portable for an incoming call.

There are other optional IEs in the {LOCATE-ACCEPT} message. One is the <DURATION> IE, which can be used to define how long the location registration and temporary identity are valid. It can be used to force portables to reattach periodically, which helps the application's mobility manager to keep its database up to date.

Rejection of the portable's {LOCATION-REQUEST} is possible, and the {LOCATE-REJECT} message also may contain a <DURATION> IE, which indicates the period during which the portable shall not reinitiate the procedure within the current location area. Rejection may occur if the system's resources are fully occupied servicing other portables or in an emergency situation, in which low-priority handsets are denied access, to guarantee service to high-priority portables.

7.6.1.2 Detach

A portable can indicate that it no longer wants to receive any incoming calls by using the detach procedure (Figure 7.12).

An MM_DETACH-req primitive from the IWU causes the portable to send a {DETACH} message to the fixed part's NWK layer MM entity, containing the portable's assigned TPUI. The fixed part's mobility application then deletes the portable from its VLR and does not attempt to deliver calls. The procedure can be used, for example, just prior to the portable finally powering itself down when the user turns it off with a "soft" on/off switch. That prevents the network from needlessly trying to deliver incoming calls to an unavailable portable.

7.6.1.3 Location Update

The location update procedure is used to inform the portable that a fixed part has updated its location areas. The fixed part sends an {MM-INFO-SUGGEST} message containing an <INFO-TYPE> IE containing "locate suggest." The portable then initiates a location request using the attach procedure. This sort of procedure should not happen very often, but it is

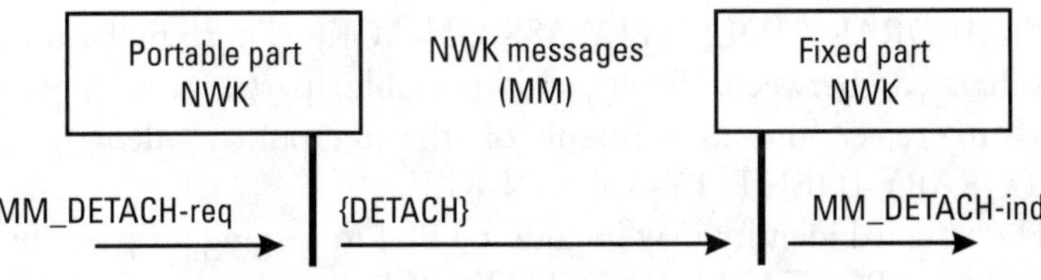

Figure 7.12 Detaching from a network. *Source:* ETSI [2].

essential if the management of the fixed system necessitates a reorganization, perhaps to reduce the sizes of the location areas when portable paging traffic increases.

7.6.2 Identity Procedures

We have already seen that a TPUI may be assigned to the portable when it attaches to a fixed system. However, the fixed part, which initiates all the identity procedures here, also may assign a portable a temporary identity at any time, not just at the time of location registration. To do that securely, the fixed part may need to know more details about the portable and its user. For that reason, an identity request procedure also is provided.

7.6.2.1 Identification of a Portable Termination and its User

A portable's formal identity is a combination of the identity of the equipment itself (the international portable equipment identity, or IPEI) and the identity of the user (the IPUI). The IPEI is a unique hardware identity embedded in a portable at manufacture. It is a 36-bit number that comprises the manufacturer's identity plus the terminal's serial number. The IPUI is a soft network-level identity intended to identify the user of the terminal. For simple residential sets, it is not necessary to identify the user separately from the terminal, so the IPUI is derived directly from the IPEI. In other applications (e.g., a PABX), there are other means of setting the IPUI separately, to identify the user as well.

The fixed system application may request the portable to identify itself at any time, by getting the IWU to send an MM_IDENTITY-req primitive to the MM entity. That causes an {IDENTITY-REQUEST} message to be sent, which may request the declaration of any combination of the portable's formal identities (IPEI and IPUI) or its assigned identity (TPUI). The declaration is returned in an {IDENTITY-REPLY} message.

7.6.2.2 Assignment of Temporary Identity to the Portable Termination

Apart from the location procedures' roles in assigning a temporary identity to the portable, a separate procedure exists to do that as well. Two messages are specified, {TEMPORARY-IDENTITY-ASSIGN} and {TEMPORARY-IDENTITY-ASSIGN-ACK}. Those messages simply are exchanged between fixed and portable parts. The portable is allowed to reject the assignment of the temporary identity with a {TEMPORARY-IDENTITY-ASSIGN-REJ}.

The assigned identity may include a TPUI for paging purposes (which is sent inside a <PORTABLE-IDENTITY> IE), a network-assigned identity (which is sent inside a <NWK-ASSIGNED-IDENTITY> IE), or both to assign the portable an identity for the application to use, which may not want to use any of the DECT-specific identities.

7.6.3 Access Rights Procedures

Access rights procedures exist to allow a portable to obtain access rights to one or more fixed systems. The procedures allow data to be loaded into a nonvolatile memory in the portable, comprising information such as the IPUI, the PARK, and other service-specific information. There also are procedures that terminate the access rights of a portable. The access rights procedures do not provide a secure procedure to allow a portable to get hold of an authentication key without disclosing it openly on the air. For that purpose, the key allocation procedures are used (covered in Section 7.6.5).

A portable can be registered as able to access many different services. Each service must have a separate data storage area within the portable, so a portable has to be built with at least one storage area for one service subscription. However, the portable may have any reasonable number. For attaching to a specific service and making and receiving calls through it, the portable has to select one of those storage areas and then operate using just that subscription information.

Think of this arrangement as creating a portable that at any one time may be one of several different telephones each subscribed to a different service or system. You have a (hypothetical) switch on the portable that has different positions, labeled "the Smith residence, John's line," "the Smith residence, Janet's line," "Janet's PABX line at the GurgleTone Corporation," "John's subscription on the Magenta public cordless service," and so on. When the portable is turned on and scans for a service it recognizes, it is the PARK in each of those storage areas that it compares with the system information on each beacon it sees. If a match is found, the portable automatically becomes one of those telephones by selecting and using the matching subscription data.

7.6.3.1 Loading Access Rights Information

Every DECT or PWT portable has a default IPUI, called an "IPUI type N," which is basically the equipment's unique 36-bit hardware serial number (the IPEI) augmented by 4 bits identifying that this particular IPUI contains the IPEI. Other types of IPUI may contain, for example, the portable's assigned telephone number plus the identity of a public or private operator that provides it telephone service under that number. Those other types of IPUI have to be assigned to the portable as part of a subscription process, and that is the purpose of this procedure. More details of DECT's system identities are given in Section 8.4.

A request for registration, or access rights, is initiated by some application-specific procedure such as a "register" function on the handset. The application's IWU sends an MM_ACCESS_RIGHTS-req primitive to the protocol stack, and the stack sets up a link and sends out an {ACCESS-RIGHTS-REQUEST} message containing a <PORTABLE-IDENTITY> IE giving the portable's default IPUI. The fixed system receives the message, and its IWU gets an MM_ACCESS_RIGHTS-ind primitive from the stack. If it is expecting such a request, it may respond with an MM_ACCESS_RIGHTS-res indicating "accept." For example, a residential base station might have a "register" button, which places it for some short time (say, 1 minute) into a state in which it will accept an attempt from a handset to obtain access rights. For a PABX, the system administrator probably has a password to get access to a subscription management feature that can be used to set it up to accept a specific handset's request for registration.

On being told that the fixed system's application has accepted the request, the MM entity sends an {ACCESS-RIGHTS-ACCEPT} message containing at least a <PORTABLE-IDENTITY> IE with the correct IPUI for the portable to use in identifying itself to the service, and a <FIXED-IDENTITY> IE with the PARK for the portable to use in the future to recognize that it has access rights to this system. Other information also may be sent, such as additional PARKs and information on the authentication procedures to be used. In addition, information may be sent to the portable that is specific to the IWU and to the application layers, either for storage or to initiate application-specific action to complete the registration process.

7.6.3.2 Terminating Access Rights

Terminating access rights is a procedure similar to obtaining access rights. It originates at the portable and is designed to erase a specific IPUI and its related information from the subscription memory of the portable and the HLR in fixed systems.

The {ACCESS-RIGHTS-TERMINATE-REQUEST} message from the portable contains the IPUI for the registration that the portable wants to erase, and the fixed system may issue an {ACCESS-RIGHTS-TERMINATE-ACCEPT} message to confirm to the portable that its subscription has been terminated. The procedure may be accompanied by the fixed system using an authentication procedure (discussed in Section 7.6.4) to verify that the portable and/or the user has the right to delete the registration in question.

The fixed part also may initiate the procedure to terminate a portable's access rights. In that case, it also is wise for the portable to authenticate the fixed part before accepting the order and erasing one of its subscriptions.

7.6.4 Authentication Procedures

The authentication procedures allow both the PT and the FT to verify that the peer termination with which it is communicating is actually the one it claims to be. The basis for that identification is to obtain proof that the opposite part has knowledge of a secret user authentication key (UAK), without the UAK being transmitted openly. The standards deal with this procedure at the NWK layer [1] and in more detail in the security features standard [4].

7.6.4.1 Authentication of a Portable Termination

Often, authentication of a portable termination is used at the beginning of a call to allow the fixed system to verify that the portable has the right, by virtue of its knowledge of the secret UAK, to make and/or receive calls via the system. However, authentication procedures also may be used within a call, for example, to allow the system to verify periodically that the radio communication channel to an already authenticated portable has not been hijacked by a different portable to make calls at another user's expense.

The receipt of an MM_AUTHENTICATE-req primitive by the MM entity causes it to send an {AUTHENTICATION-REQUEST} message. The message identifies one authentication algorithm, of which the DECT standard authentication algorithm (DSAA) is one. This particular algorithm is required in GAP systems. The DSAA challenge contains a 64-bit random number, <RAND>, and another 64-bit parameter called <RS>, which is provided by the home network of the portable. The portable contains a standard cryptographic one-way function that computes a response, <RES>, from the <RAND>, the <RS>, and the portable's secret UAK, which is sent over the air in an {AUTHENTICATION-REPLY} message to the fixed part. The fixed part then computes the expected value of <RES> from the challenge's two parameters and the UAK it thinks the portable should know. If the expected and received values of <RES> match, the fixed part may assume the portable knows the value of UAK and is thereby authenticated.

The use of the one-way cryptographic function means that an eavesdropper, even knowing the challenge and response parameters for the most recent authentication event and perhaps many others with different values of <RAND> and <RS>, will find it impossible in a reasonable time to compute the value of UAK held by the portable and known to the fixed part.

7.6.4.2 Authentication of a User

Authentication of a user is rather like the authentication of a portable. The main differences are that the user has to enter a user personal identity (UPI) when authentication is requested, which becomes part of the authentication key. In that way, the fact that the portable knows part of the authentication key and the user knows the rest can be simultaneously verified. Of course, the procedure has to have a longer no-response timeout to allow the user to key in the UPI.

7.6.5 Key Allocation Procedures

Sometimes a portable needs to be sent information that is securely hidden from eavesdroppers. That may need to be done by a fixed system that does not support encryption of the cordless link. For example, a UAK may need to be sent for the portable to use later on when it is challenged to authenticate itself (see Section 7.6.4). The security of the authentication procedure depends on nobody except the fixed and portable systems having knowledge of the key.

The preferred means of providing the information is via some secure route such as the postal mail service, where the user enters the UAK by hand and then destroys the paper record. It also can be provided inside a DECT-specific smart card (a DAM [5]) to be physically inserted in the portable. Because the key is long (128 bits), it usually is better to avoid hand-entry procedures, even those with checksums to detect user keying errors.

Where other secure methods are not available or not suitable, the over-the-air key allocation procedure provides an alternative that derives a candidate 128-bit UAK from a preknown short (perhaps 32-bit) secret authentication code that may be more readily entered into the portable by the user who perhaps has been given the value in writing or verbally. Experience shows that keying in any unfamiliar information of more than about 32 bits is prone to user error.

The key allocation procedure uses a variation on the authentication procedure to derive cryptographically an identical candidate 128-bit UAK at both ends of the link, while simultaneously verifying that the short secret authentication code is known at both ends. The procedure provides for the derived key to be marked as unconfirmed until a later authentication process succeeds when using the derived key. The fact that the key is derived from a shorter secret number means that the derived UAK may be more readily susceptible to computation. However, the procedure is carried out only once, so the information presented over the air is much less than could be gained by watching repeated authentication challenges based on the 128-bit secret UAK. For that reason, it is more secure than it may seem from just the length of the short secret authentication code.

References

1. ETSI, *Digital Enhanced Cordless Telecommunications (DECT); Common Interface (CI); Part 5: Network (NWK) Layer,* ETS 300 175-5, Sophia Antipolis, France, September 1996.

2. ETSI, *Digital Enhanced Cordless Telecommunications (DECT); Generic Access Profile (GAP),* EN 300 444, Sophia Antipolis, France, August 1997.

3. ETSI, *Draft Integrated Services Digital Network (ISDN); Generic Functional Protocol for the Support of Supplementary Services; Digital Subscriber Signalling System No. 1 (DSS1) Protocol; Part 1: Protocol Specification,* EN 300 196-1, Sophia Antipolis, France, April 1997.

4. ETSI, *Digital Enhanced Cordless Telecommunications (DECT); Common Interface (CI); Part 7: Security Features,* ETS 300 175-7, Sophia Antipolis, France, September 1997.

5. ETSI, *DECT Authentication Module (DAM),* ETS 300 331, Sophia Antipolis, France, November 1995.

Part III
Advanced Features and Applications

8

Advanced DECT/PWT Applications

8.1 Introduction

The preceding chapters have covered the basic protocols in considerable detail, including their operation and standards. This chapter looks at some of DECT and PWT's more advanced applications and features, although not in much depth. The reader is referred to the application-specific standards and reports for more details. The documents are listed in their entirety in the bibliography.

These applications are covered here:

- Cordless PABXs;
- Interworking of cordless equipment with the ISDN;
- Data services;
- RLL applications, including repeaters;
- Interworking of cordless equipment with GSM-type networks;
- Roaming of cordless terminals across public and private networks.

Two other specific features essential to cordless systems (and particularly PABXs) also are discussed:

- System planning and deployment;
- The use of identities and addressing.

In the technical descriptions of the DECT and PWT systems and their standards, the base standards [1,2] are a menu of features and procedures covering all sorts of applications; for any specific application you need implement only a subset. For many applications, there are application-specific profiles, which are other standards describing which of the base standard's features are required. All profiles which support speech telephony, are based on the GAP [3] so that DECT handsets conforming to those profiles also are GAP compliant. If the application does not include speech service, the European regulatory regime does not require DECT systems to comply with any of the application-specific standards.

8.1.1 Basic Voice

For the support of all voice applications, as covered in detail earlier in this book, the essential subset of DECT features is the GAP in Europe [3] and the CPAP in the United States [4]. In Europe, all DECT systems that are capable of supporting the "3.1 kHz telephony" speech service must comply with the GAP. That means DECT handsets may be used to get speech service from any base station that offers it, regardless of its manufacturer. That is not a requirement in the United States, but the CPAP gives the opportunity to ensure interoperation.

Because of the 10-ms TDMA frame, digital streams such as speech cannot be delayed less than that in passing through a DECT or PWT system. In practice, another 1 to 2 ms also is incurred in any analog circuit that incorporates filtering, so the standards permit a total one-way speech delay of up to almost 14 ms. As a result, all fixed parts have to incorporate some echo control. The precise nature of the echo control is covered in the speech coding and transmission standard [5].

8.1.2 Cordless PABX

The cordless PABX probably is, after the residential telephone, the main application for which DECT and PWT originally were designed. It certainly drove the initial phases of protocol design, leading to fundamental choices in the PHL and the MAC layer, that give the system its characteristics, provide adequate flexibility, and ensure the capacity to support cordless offices of many thousands of users.

The cordless PABX changes one of the old but fundamental assumptions of a wired PABX that a dialed number corresponds to one (or more) fixed ports at its periphery. There are two ways of making that change. The first is to build into the PABX's software the new paradigm that a terminal may be connected to any "port" (i.e., attached to any radio base station) and that it may change ports at any time (even during an established call). The second way is to add an adjunct to a standard PABX that transparently routes calls from fixed ports to the appropriate radio cell. Both approaches are covered in Section 8.2.

Two key topics related to cordless PABXs and other multicell systems also are covered: system planning and deployment in Section 8.3 and system identities in Section 8.4.

8.1.3 ISDN Interworking

The connection of DECT and PWT systems to the digital public telephony network, the ISDN, is not in itself considered to be ISDN interworking, since the protocol is designed to be generic and not network specific. However, handsets are allowed to (1) access ISDN-specific services in such a way that the handset plus base station look like an ISDN terminal; or (2) carry an ISDN connection so that a normal ISDN terminal can be connected to a socket on the handset. Those two means of interworking with the ISDN are covered, respectively, by two different ISDN interworking profiles, which are described in more detail in Section 8.5:

- ISDN access profile (IAP) [6];
- ISDN intermediate profile (IIP) [7].

8.1.4 Data Services

The protocol creates bidirectional 32-kbps continuous circuits to carry coded speech. For creating simple data streams, that channel also can be used as an uncorrected data connection, or it can be used with error control to deliver data circuits with lower data rates (the DLC's LU5 and LU6 services). However, the base standards also support more versatile packet-mode communications that are used to build up a wide-ranging set of data capabilities. The generic data transfer capabilities in the base standards are brought together in the data service profiles, into capability sets that support specific applications and services. The profiles are labeled A to F and subdivided into class 1 (without mobility) and class 2 (with mobility) (Table 8.1). More details of these services are contained in Section 8.6.

Table 8.1
Summary of Data Service Profiles

Profile	Classes	Applications
A/B	1, 2	A generic frame relay service. For example, to carry connectionless traffic up to 24 kbps (type A) or 552 kbps (type B), extending token ring and Ethernet LANs. Class 2 covers mobile applications, including direct interworking with the Internet.
C	1, 2	High-integrity generic data stream services built on the generic frame relay service of type A/B, including interworking at V.24 interfaces.
D	1, 2	Isochronous transparent transport of synchronous data streams for closed user groups (class 1) or for public and private roaming applications (class 2).
E	2	A roaming low-rate short message service (e.g., alphanumeric paging). Point-to-point and point-to-multipoint services are supported.
F	2	A mobile multimedia messaging service including e-mail, facsimile, short-message service (SMS), and world-wide web (WWW) access, including file transfer protocol (FTP) and hypertext transfer protocol (HTTP).
PPP	2	Support for Internet protocol (and other datagram types) over PPP, based on the A/B class 2 and C class 2 profiles.

8.1.5 Radio Local Loop

One of the more important areas in the application of radio in the public telecommunications service is the replacement of the copper wires from a local telephone exchange into a home or other building. The rationale for doing that by radio is that the cost of installing and maintaining copper cables is very high, and although radio systems initially may be more expensive, they ultimately work out cheaper to install and maintain.

The protocol's flexibility in carrying many different telecommunications services makes it ideal as a relatively short-range radio replacement for the "local loop." It provides both traditional local-loop services and advanced features. In trials using high-gain antennas, ranges of up to 5 km have been achieved. In practice, in real environments with obstructing buildings and trees, for example, more modest ranges may be achieved on average.

The RLL access profile is divided into two parts:

- Part 1 covers basic services, including standard telephony, a 64-kbps bearer service, and the essential operations administration and maintenance procedures needed for an RLL.
- Part 2 adds ISDN and broadband packet data services.

Section 8.7 gives more details, and Section 8.8 gives details of the wireless relay stations that allow all the protocol's services, including RLL, to be extended in range where obstructions or other problems prevent direct service from being provided from a base station.

8.1.6 DECT/GSM Interworking

DECT and PWT basically are radio access technologies that do not specify any network mobility functions. However, they do specify support for mobility functions existing outside the cordless air interface. That makes DECT or PWT an ideal partner for a digital cellular technology, such as GSM, that already specifies how to handle mobility within its network. In fact, there are not always perfect matches between the models that the cordless system has for things such as mobility, identities, and security, so the partnership between the cordless access technology and the cellular mobility infrastructure is not always ideal. Nevertheless, there are two different ways to make the necessary adaptations to allow cordless systems to be connected to GSM900, DCS1800, or PCS1900 infrastructure: (1) connect cordless fixed parts to the A-interfaces of a GSM network or (2) connect complete cordless PBXs to GSM infrastructure through the ISDN using the DSS1+ protocol [8]. Both mechanisms have application profiles that allow GSM's mobility and other functions to be made available to cordless handsets. Users then can have the same services through their cordless handsets when they are out of range of their GSM system.

There is also a third means for cordless and cellular systems to interwork: having a dual-mode cordless and cellular handset. That does not require the interworking of fixed systems, and the handset just selects the appropriate operating mode for its environment, for example, DECT when it can see a DECT system and GSM otherwise. More complex and useful arrangements also are available. The interworking of cordless and cellular systems is covered in more detail in Section 8.9.

8.1.7 Public/Private Network Cordless Terminal Mobility

Cordless terminal mobility (CTM) is described in DECT's CTM Access Profile (CAP), which details how to allow users of cordless terminals to roam between multiple interconnected cordless networks and receive their telephone service. It applies whether the relevant networks are public or private. The CAP provides the means by which any cordless network can communicate with another to allow it to serve handsets that have their homes on other networks.

The ability to roam between networks also may be provided by interconnecting the cordless networks via the GSM network (using the DECT/GSM interworking profile). The CAP provides alternative mobility functions through the ISDN. Section 8.10 gives more details.

8.2 Cordless PABX

Strictly speaking, the DECT and PWT standards have nothing to say about either cordless PABXs or their private networking. They are telecommunications access protocols. However, built into the cordless protocol are the support mechanisms for cordless PABXs and the means to support most types of communications network, including private (or enterprise) networks. Where voice is carried by a DECT PABX, it can be assumed that any unrecognized handset is GAP compliant. If it is not, then interoperation cannot be guaranteed. If data service is provided, one or more of the data profiles, according to the service, also may be supported (covered in Section 8.6).

8.2.1 What is a Cordless PABX?

If you take a normal PABX and attach a lot of normal DECT or PWT cordless telephones to a number of the ports, do you have a cordless PABX? No, of course not. What you have is a limited form of mobility for the users, centered on the fixed sockets into which they normally would plug a wired telephone. What you effectively get is an extension of the telephone cord between the handset and the base station. That in itself may be a useful step forward, but it is not a cordless PABX.

The cordless PABX goes further. The ideal cordless PABX effectively replicates the fixed socket you use for your normal telephone *everywhere* within its radio coverage area. That requires a shift of emphasis in the central switch serving the cordless users. Normally a dialed number is translated by the switch's central processor to a fixed location. In that sense, the location of a numbered port is at the periphery of the switch. An incoming call is routed to the port corresponding to the dialed number. But now the periphery of the switch is connected to a group of cordless base stations instead, and there is no longer any correspondence between the port through which an incoming call must be switched and the number dialed for the telephone call.

That means that the cordless PABX has to maintain a database that connects a dialed number to a particular portable set rather than a fixed port, and a second database that records at which location that set is to be found. Sometimes the second database can be imprecise. In the DECT/PWT system, a call to a handset is put out over at least one cluster, which may comprise as many as 255 radio cells; and the handset is responsible for maintaining knowledge at all times about the best cell to contact within the cluster whenever it hears the broadcast information notifying it of an incoming call. That is one of the main mobility design features of the protocol. By decentralizing some of the mobility management to the handsets, the mobility management in the central cordless switch is greatly simplified for even very large systems.

Supporting both mobile and nonmobile terminals is easy. Fixed telephone sets just are treated as mobile handsets that never move.

8.2.2 Differences Between a Cordless PABX and a Wired PABX

It may seem obvious, but a cordless PABX has to be sold in a very different way than a wired PABX. The reason is that the first and most important service it provides is radio coverage. By the time customers have successfully operated their very first handsets, they will have had to install and pay for all the base stations needed to provide radio coverage. Each radio base station will have cost substantially as much as a reasonably featured business telephone set, so the cost per line of the first few handsets will seem enormous.

Yet telephone system buyers—and their senior management—probably will be used to dimensioning telephone system purchases by the number of users rather than the area of the premises, and they will be used to buying their switching systems at a certain cost per line. So selling mobility in the workplace is very different from selling a normal fixed PABX, and the potential difficulties in selling the advantages of in-building mobility and changing the outlook of telephone service buyers should not be underestimated.

8.2.3 Integrated Support Versus Adjunct Support

Changing one of the basic switch paradigms from "one dialed number corresponds to one physical port" to "one dialed number corresponds to one handset anywhere in the radio coverage area" usually means a fundamental change in a PABX's core switching software. Making that change can be done in one of two ways. The first way is to bite the bullet and make the changes in the core software; the second is to connect a special-purpose cordless switch to the main PABX as an adjunct, and allow the adjunct to translate from one paradigm to the other. If you are providing mobility for a large PABX, it may be far easier to connect an adjunct mobility processor to the switch rather than update the core software to add mobility support.

The normal way to design an adjunct is to realize that the normal telephone services you are getting should come from the core switch. The adjunct should just add mobility. The adjunct should then be programmed, in combination with the portable, to pass through any request for special features from the portable and any responses in the opposite direction.

8.2.4 Nonconcentrating Adjunct Versus Concentrating

There are perhaps two main types of cordless PABX system you could design: a concentrating system or a nonconcentrating system. Fully integrated cordless support within a PABX normally will be of the concentrating variety, whereas an adjunct reasonably may be of either type. We illustrate the difference here by referring to adjunct support for cordless mobility (Figure 8.1).

The key to the nonconcentrating adjunct is that the number of handsets is equal to the number of trunks ($Nh = Nt$). The adjunct simply associates one handset with one trunk. Hence, an incoming call indication on one trunk tells the adjunct which handset to call. The number of cells is not related to the number of trunks or handsets but rather to the coverage area required.

In this sort of system, a handset may be twinned with a fixed phone, so it will ring at the same time. In that way, it is possible to introduce a mobility service in parallel with an existing fixed system. The features of the fixed system are available via the existing fixed phones, and users have the ability to receive and originate calls when away from their normal place of work.

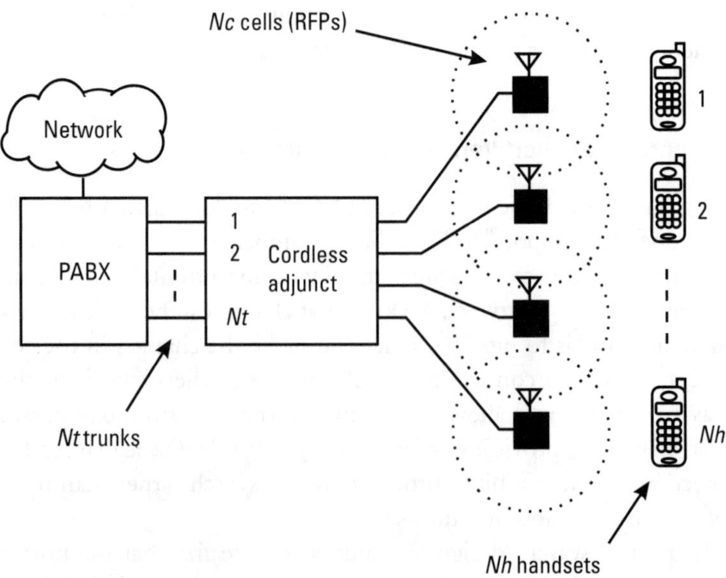

Figure 8.1 Adjunct support for cordless mobility in a PABX.

This sort of behavior is possible even with analog trunks between the adjunct and the PABX. Simply detecting the ringing voltage on the trunk is enough for the adjunct to initiate a call to a DECT or PWT handset. That way, a cordless adjunct can connect to any PABX with analog extensions, regardless of who manufactures either the PABX or the adjunct.

There are some disadvantages. With analog interconnect, the ringing voltage will be cadenced, with gaps between the ringing voltage pulses. The adjunct has to implement a timeout long enough to cover the gaps and must continue to cause the handset to ring across those gaps. When the incoming ringing stops, it may be some time before the handset stops indicating an incoming call. That may cause a handset user to answer a call attempt that has just ceased. Also, where the adjunct provides for a cordless handset to act as an additional instrument in parallel, the handset may continue to ring for a short period after a fixed extension has been answered.

In a nonconcentrating adjunct, special features seen by the user usually are those of the PABX rather than the adjunct. The features of the fixed PABX are made available via the portable phone if they can be accessed via special sequences of the normal keys 0 to 9, *, and #. The keys may be dialed by the handset user directly or be assigned within the adjunct to special key functions on the handset and translated in the adjunct itself. The nonconcentrating adjunct usually also provides no internal switching. That is left to the main PABX, and all calls are passed to the main switch even if they are between cordless handsets supported by the adjunct.

In the concentrating adjunct, there are fewer incoming trunks than handsets ($Nt < Nh$). In that case, an incoming call must be accompanied by a signal that indicates which of the handsets it is for so the adjunct can make the appropriate connection. Communication of that type is largely the province of digital trunks and is usually an area where the digital trunking is proprietary to the PABX. In that case, the adjunct may well not be a general-purpose adjunct and may be specially designed to work with a specific switch.

8.3 System Planning and Deployment

The first thing to say about the planning and deployment of a multicell public DECT or PWT system (or any cordless PABX) may not appear to be directly related to planning but rather to the marketing of the system. It bears repeating that you need to sell a customer radio coverage before you even sell the first terminal. That principle is the key to the deployment of a cordless system as well. You first must plan the coverage area of a system before you even consider the numbers of users and terminals the system will support.

Planning and deployment as we describe them here do not really apply to single-cell systems, only perhaps the capacity calculations. This treatment therefore mostly applies to multicell systems, such as PABXs and public access systems.

8.3.1 Radio Coverage

DECT and PWT systems may seem like cellular radio systems in miniature. They are indeed similar. However, unlike normal cellular radio systems, the great thing about the protocol's internal dynamic channel allocation function is that you never need to be concerned about allocating frequencies to cells. Provided your equipment follows the rules for channel selection within the standard and is not overloaded with users, the protocol will take care of all aspects of frequency planning.

The placement of cells to provide coverage is not an exact science. Inside buildings, it is even worse than outdoors. The type of office may vary from one end of the scale, the more traditional offices-and-corridors building with steel-reinforced concrete floors, to the other, a large open-plan office covering the entire floor of a building. In the former case, the radio attenuation between offices and between floors probably will be so great that most radio cells may have no more than a 20m radius. In the latter case, a single strategically sited cell may be able to maintain radio coverage of an entire floor of the building and perhaps the floors above and below as well. For outdoor public systems, you can find equally extreme situations.

Environment-related variable attenuation is not the only problem with getting adequate radio coverage. Radio propagation at the DECT and PWT frequencies also is subject to multipath fading, where multiple radio reflections add up in or out of phase to increase or decrease the radio signal at any one point (Figure 8.2). That leads to unexpected local dropouts in which the radio signal falls below the sensitivity limit of the radio receiver.

These unexpected dropouts lead to the actual radio range being somewhat less than the range anticipated from the receiver's measured sensitivity and the ideal free-space signal-loss curve. For that reason, a fading margin usually is applied in radio coverage planning, which is, in effect, a reduction of the actual sensitivity to an apparent sensitivity that covers the dropouts. That prevents the appearance of pockets of poor or absent radio coverage.

8.3.2 Cell-Site Planning

Because of the extreme variations in radio environments, planning the placement of RFPs is not an exact science. The "kitchen-planner" approach with a

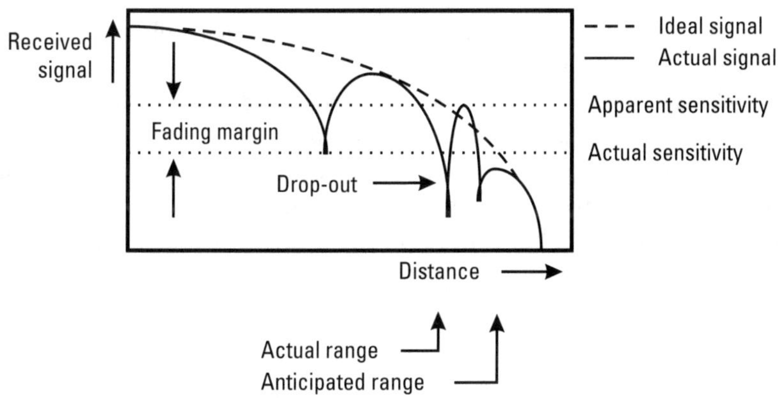

Figure 8.2 Fading margin.

plan of the building or the district and a categorization of the radio environment on a scale ranging from "easy" to "difficult" can be made to work but often leaves too many radio coverage holes, especially if you are trying to be reasonably economic with the number of cell sites. Remember that each cell site represents a significant investment in the following:

- Installation manpower;
- Equipment at the cell site;
- Wiring from the cell to the central controller;
- The cell's share of equipment at the central controller;
- The cell's power supply (either local or via the wiring from the central controller).

Compared to the kitchen-planner approach, a better installation usually is achieved with a field deployment tool that measures the actual radio coverage from a portable RFP placed at a candidate cell site. The limit of coverage is where the signal from the portable RFP drops below the terminal's apparent sensitivity, as experienced by a user. To determine that, you can start with the –86 dBm figure required of a GAP-compliant handset (–90 dBm for PWT), but you must allow for multipath fading (perhaps 12 to 18 dB) and some overlap between cell edges (about 3 dB is probably reasonable if the system has seamless handover and the extra handover traffic is acceptable; 6 dB otherwise). The trade-off here is that the engineering manpower required to do such a survey is greater than that needed with just a building plan, an installation guide, and a pencil.

RF safety is another matter you need to consider when planning cell sites. Unlike a normal portable, which may transmit on normal power (250 mW) for just one timeslot every 24, the RFP may be communicating with several handsets and may also have a directional antenna. In most normal cases, safety does not present a problem, but annex A of the DECT PHL standard [9] does give guidance on safety distances.

8.3.3 System Capacity

With a single cell, you may have up to 12 duplex radio channels available per RFP. Cheaper RFPs with perhaps a slower RF synthesizer may be able to use only six. Other limitations may apply, such as the connection between the RFP and the controller, which may have to transport data a couple of kilometers in big systems. For example, it may be economic to provide only backhaul capacity for up to eight duplex radio channels over long signaling links. Regardless of the limiting factor, there will be a limit to the maximum number of duplex radio channels that can be handled by a single RFP.

Providing coverage in a cell requires the deployment of at least one RFP, but capacity considerations may then dictate that more than one RFP be sited within that cell. An RFP with a 12-channel capacity and a 12-channel backhaul link will provide 5.3 Erlangs of capacity (Table 8.2) at a 0.5% grade of service (GoS), that is, there is a probability of 0.5% that any particular call cannot be made because no radio channels are available. That particular number can be looked up in the Erlang B tables under the 12 trunks column and then locating the capacity for a 0.5% GoS figure. The better the GoS (i.e., the smaller the GoS number), the smaller the capacity in Erlangs available from the 12 channels.

It is typical to plan for office systems on the need for about 0.15E of traffic generated by a user's terminal. Roughly speaking 0.15E means that the user

Table 8.2
Number of Users Served Versus Available Base Station Channels

Available Channels	Total Capacity at 0.5% GoS (Erlangs)	Users at 0.05 Erlangs	Users at 0.10 Erlangs	Users at 0.15 Erlangs
12	5.3	106	53	35
8	2.7	55	27	18
6	1.6	32	16	11

of the terminal spends 15% of the time on the telephone. Experience shows that reality sometimes can vary greatly from that "average" and a real deployment situation will have to be carefully assessed.

However, at 0.15E per user, the 5.3E capacity of a 12-channel RFP can serve 5.3/0.15, or about 35 users. If the RFP has a slow synthesizer and can use only 6 channels instead of 12, the capacity drops more than you might expect, supporting only 11 users. In those sorts of environment, it may be more economic to use better quality synthesizers in the RFPs than to deploy colocated RFPs to get the required capacity.

In a big open-plan office, a cell diameter may reach 100m (an area of about 8,000 m^2), perhaps more. The number of potential users within that area easily could exceed 35 (at 20 m^2 per user in an office full of desks, there might be 400). If so, the number of channels in the cell has to be increased, or the size of the cell has to be decreased and more cells put in to provide the service.

Often, increasing the number of RFPs within a cell is easy. First, there is a site at which the original cell is located. That usually will be able to accommodate a second RFP, and the wiring to the controller has to go to only one spot. Also, if external power has to be supplied to the RFP (rather than through the telecommunications leads), there usually is power available already. However, reducing the size of a cell usually is the best way to increase capacity. That is because at any point the signal strength available from a cell usually is greater, since you are nearer the center of the cell. Also, it means that the capacity of neighboring cells may be available to users within a cell in the event the current cell's capacity ever does get used up completely.

There is a disadvantage to shrinking cells in that the radio interference background increases due to the proximity of the other cell sites. Nevertheless, up to a point, where the cell may be as small as 15m in radius, the system's channel allocation algorithm will cope with finding channels to use that are sufficiently clear of interference. However, the disadvantage of splitting the cell is that more sites have to be found with wiring back to the main controller and perhaps extra power supplies. The cost of installing a power supply at any point may add significantly to the cost of installing the cell, so line powering of the base stations often is an attractive option when overall installation costs are taken into account.

8.3.4 Incoming Calls (Paging)

When planning a big cordless PABX or any other high-capacity system, you may need to think about incoming-call traffic and whether to divide the system into multiple clusters (see Section 5.2.6).

Paging traffic is carried simultaneously on the beacons of all cells within a cluster (see Section 5.5.6). That means you normally get the capacity to issue just six pages per 160-ms multiframe (see table 5.8), or 37.5 paging messages per second. The pages may have to be repeated if the handset is not responding or is missing the message for some reason, so the real capacity is somewhat lower. Fortunately, only a reasonably big system with a lot of heavy users may have to handle as many as 10 incoming calls per second during the busiest periods. Nevertheless, if your system has to handle anywhere near that level of call arrival, you need to divide it into clusters and assign separate LAL identities to split up the paging load.

8.4 Identities, Addressing, and Security

The identities of portable and fixed systems are things that a system manufacturer, installer, or user has to manage for a DECT or PWT system to function properly and with security.

The identities of complete fixed systems (see also Section 8.4.3) are called access rights identities (ARIs) because they are used by portables to determine if they have access rights to the system. In simple cases (such as residential telephones and very small PABXs), those identities are fixed and set up by the manufacturer. In more complex or bigger systems (e.g., bigger PABXs, interconnected systems, or public access systems), the identities are loaded into the system by the installer or operator. In all cases, the identities are then broadcast by the fixed system on its beacon. It is important to note that when systems are loaded with identities, they really are the identities of the service offered, not necessarily the system offering it. It is entirely possible to load the same sets of identities into two completely different systems, provided they both offer the same service.

Complete system identities are subdivided so that individual cells (RFPs) within the system may be identified. That is done by allocating each RFP an RFP Number (RPN). That also allows groups of cells to be identified together as location areas, which may be used to manage the way in which portables are tracked as they move.

When a portable is given access to a fixed system, the fixed system's ARI is given to the portable. The portable stores the ARI for future reference, in which form it is called a PARK. The portable also is told how many bits of the PARK must be checked against a broadcast ARI to determine if the portable has access rights to the system. If the portable later hears a system broadcasting an ARI, it can compare the ARI with any PARKs it has stored. If there is a match to at least the specified number of bits specified by the PARK length indicator (PLI), the portable knows that it will be able to gain access to the system and can then attach to the system to make sure the system knows it is there.

The identity of a portable itself may be one of two types. The first is the unique hardware serial number of the portable, the IPEI (international portable equipment identity). It is possible for a fixed system to regulate access based on the portables' IPEIs, but then only certain pieces of equipment, rather than certain users, are allowed access a fixed part's services. This is not sufficient for all applications. It is therefore possible to assign an identity to the portable (when it is given access to a system) that is there to identify the portable's user rather than the hardware. Those kinds of identity are the IPUIs (international portable user identities). The portable may be given an IPUI at the same time the portable is given the fixed system's PARK and PLI. If the identities are stored inside a DECT smart card (a DECT authentication module), the user may transfer an identity from portable to portable. If the user travels and uses different equipment in different locations or if equipment has to be replaced, the user's identity remains and the service continues.

The portable's complete identity (as far as accessing a given service is concerned) is a combination of its stored IPUI + PARK + PLI. However, a portable also may be given a TPUI (temporary portable user identity) when it successfully attaches to a fixed system after recognizing the fixed system's ARI. The TPUI is used to save bandwidth when the portable is paged as a result of an incoming call. Paging messages can address a portable by its IPUI, but the TPUI is shorter and uses up less bandwidth. This state of affairs is illustrated in Figure 8.3.

8.4.1 Portable-Hardware Identities

Every portable is unique. They are each assigned a 36-bit hardware identity at manufacture that cannot (or should not) be changed during the lifetime of the

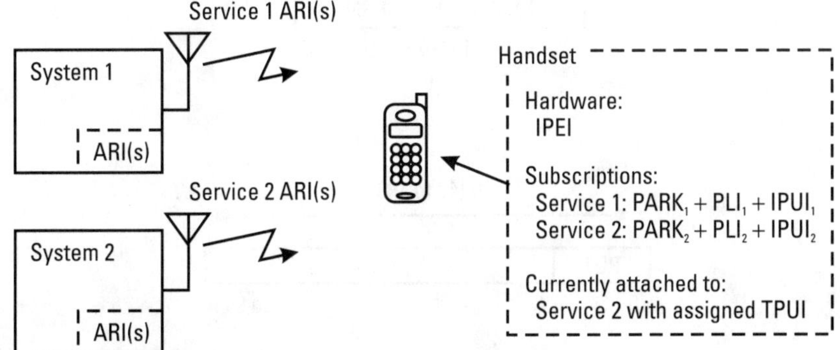

Figure 8.3 Storage of system identities.

hardware. That identity is the IPEI, as shown in Figure 8.4(a). It has a standard textual representation for printing on a label or for displaying on the portable's (or any other) display device.

The first 16 bits of the IPEI, the equipment manufacturer code (EMC), is allocated by a standards body to be globally distinct for each manufacturer. In the case of European manufacturers, an EMC may be obtained from ETSI (see the glossary for contact details). For the United States, ETSI also acts to allocate those identities.

The remaining 20 bits allow a manufacturer to allocate a unique serial number to each of just over one million handsets or other portable terminals. A company that makes more equipment than that needs to get another EMC. A manufacturer should never rely on identifying its own equipment by the first EMC it is allocated.

8.4.2 Assigned Portable Identities

As well as having a fixed electronic identity, a portable also is assigned an identity to use when it is registered to use a service. The identities must be stored in the portable for future use in accessing services. The assigned identity is keyed into the portable or downloaded via an on-air subscription process.

A single portable may have many different assigned identities, one given to it by each service with which it is a registered user with access rights. Actually, it is not really the portable equipment that is assigned the identity when registration takes place but the portable's user; that identity is the IPUI, as illustrated in Figure 8.4(b), which may take several different forms depending on its use.

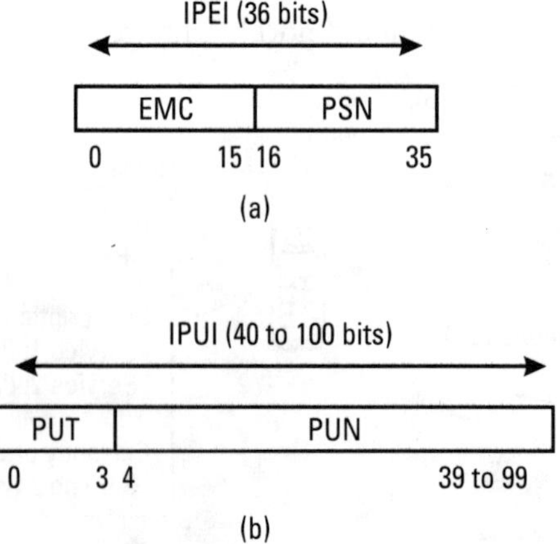

Figure 8.4 (a) The portable's IPEI; (b) the portable's IPUI.

It is possible that the assigned IPUI is stored in the portable in a memory that is not separable from the equipment. In that case, the user and the equipment are not separable. However, with the use of a smart card with DECT-specific storage, a DAM, it is possible for a user's identity to follow that user from one handset to another. The different IPUIs are listed in Table 8.3.

A portable that is otherwise unregistered uses the default IPUI of type N. For residential systems, IPUI type N may be the only IPUI the portable ever has.

8.4.3 Fixed-System Identities

Every fixed-part radio in a DECT system broadcasts a beacon on which is carried one or more fixed-system identities (ARIs). The broadcast mechanism that allows that is described as a vertical thread of procedures and data structures in

Table 8.3
International Portable User Identities

IPUI Type	Use
N	Residential systems and default use (for small systems) where the hardware identity, the 36-bit IPEI, is enough to identify the user.
S	General PSTN/ISDN use, where the identity is a 60-bit binary-coded decimal PSTN, or ISDN telephone number.
O	Standalone private system where the identity is a binary-coded decimal telephone number (60 bits), possibly significant only to the local network.
T	Extended private system where roaming is supported over several private networks. The identity is a combination of the 16-bit equipment installer code (EIC) and a 44-bit binary-coded decimal telephone number, possibly significant only to the local network.
P	Public or public access system (including RLL) where the identity is a combination of a 16-bit public operator's code (POC) and an 80-bit binary-coded decimal account number.
Q	General public access system where the identity is an 80-bit binary-coded decimal bank account number to which charges for service are made.
U	General public access system where the identity is an 80-bit binary-coded decimal credit card account number to which charges for service are made.
R	Public GSM-connected system where the identity is the subscriber's GSM identity, a 60-bit international mobile subscriber identity (IMSI).

Chapters 3 through 7. One of the broadcast identities is selected by the system installer or operator as the system's primary ARI, PARI, which is broadcast the most frequently. The system is allowed to broadcast secondary ARIs, SARIs, which are broadcast less often than the PARI. You might, for example, want to broadcast a SARI if you have an arrangement with a neighboring company to allow their employees to use your PABX. In that case, their PARI would be programmed into your system and broadcast as a SARI. It is also possible to program tertiary ARIs, TARIs, which are not broadcast but available only when a portable asks for the system's TARIs.

The fixed part's ARIs are derived from some basic information about the fixed system, its purpose, and its capabilities. The main issue is that the manufacturer of a system has to decide into which class its equipment falls. That is shown by the first three bits of the ARI (Figure 8.5). The remaining bits of the ARI vary according to the ARI class.

Broadly (but not exclusively), systems fit into up to eight classes, of which six currently are allocated as shown in Table 8.4. A residential single-cell cordless telephone set, for example, is best classified under class A. Part 6 of the DECT standard, "Identities and Addressing," [10] shows that ARI class A comprises three bits identifying the ARI as a class A ARI (the access rights class or ARC) plus 16 bits called the equipment manufacturer's code (EMC), which a manufacturer must obtain from ETSI, plus 17 bits called FPN (fixed part number), which a manufacturer must allocate uniquely (avoiding FPN = 0) to provide a unique hardware-related ARI.

8.4.4 Identities for Location Areas Within a Fixed System

Many cordless systems are small enough not to need any more than one identity for the entire system. Sometimes, however, you do need to distinguish one part of the system from another. To do this, the fixed system's identities are augmented by taking the PARI, tagging on to it the identity of the cell in which it is being broadcast, and adding one more bit to say whether any SARIs are

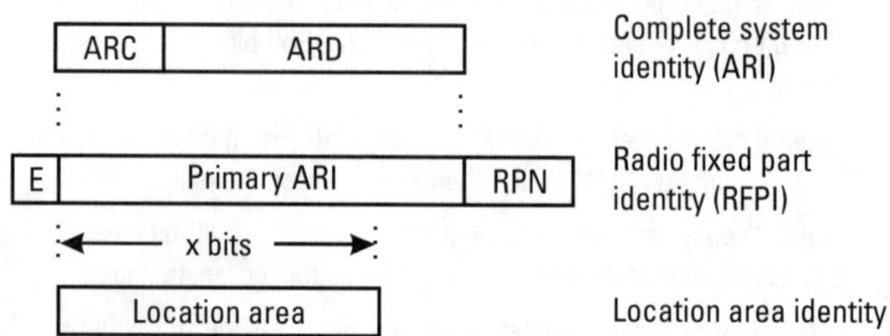

Figure 8.5 The fixed system's identities.

Table 8.4
Fixed-System Identities (ARIs)

Class	System	SARI or TARI?
A	Residential or small private PABX, from one to seven cells. Identities are preconfigured and based on fixed hardware identities, which change if any equipment is replaced. The fixed identity contains an EMC, which must be obtained from ETSI.	No
B	Larger multicell private PABX, where the identities are programmed. The identities include an ETSI-assigned EIC as well as an EMC.	Yes
C	Public telephony systems. The identity contains a POC, which is assigned by ETSI.	Yes
D	For public GSM-system operators who allow their subscribers to use DECT to access their networks (see Section 8.9).	Yes
E	Portable-to-portable direct communications (private).	Yes

available for the system (see Figure 8.5). That creates the RFPI, which identifies a specific cell within the system.

If it is important to define different location areas within a system to manage portable mobility, the portable is told, when it attaches to a system, the number of bits from the PARI to look at to determine which location area it is in. That allows location areas to comprise a number of cells, depending on how many bits of the PARI are assigned to the location area.

Whenever the portable strays outside its current location area, which can be determined by looking at the system's RFPI as broadcast from the best cell, it is required to perform a location update to inform the system of its new location area.

8.4.5 Authentication

The authentication of a DECT or PWT system is based on establishing that the system has knowledge of a secret key without the key being transmitted over the radio interface. Just how the key gets to the system without anyone else seeing it is not primarily the role of the protocols, although it is critical to the real security of the authentication. However, verifying the key without revealing it is a part of the standards.

In principle, authentication of a handset can occur at the start of any call and also can take place regularly within that call. The first authentication establishes the bona-fides of the handset; later authentications establish that the link has not been taken over by an intruder.

The basic principle for authentication of a handset (Figure 8.6) is that the secret key is held in secure storage. The base station knows the key, but instead of asking for the key to be sent in-clear, it "challenges" the handset to authenticate by sending it a random number. The authentication key is used to encrypt the random challenge, and the result is sent back to the base station.

The base station, knowing the random challenge and the secret key, performs the same action as the handset and compares the handset's response with its expected response. If the two match, the base station may reasonably assume that the handset knows the secret key.

Although the algorithm for encrypting the challenge with the secret key is not a secret, it is constructed carefully so that knowing the response and the random challenge does not give any clue about the key itself. Calculating the key knowing the challenge and the response is just too computationally expensive to work. Similarly, since there are many possible different responses, in principle a different one for each of the many possible challenges, it is no good listening to the responses and trying to respond with one you already have heard. The same challenge will never be repeated in any reasonable time.

The algorithm, the DSAA, is not a secret. However, it is not revealed to anyone without their signing a nondisclosure agreement to limit the algorithm's exposure to malicious cracking attempts. It may be obtained by contacting ETSI, which will indicate the current arrangements for obtaining the documentation.

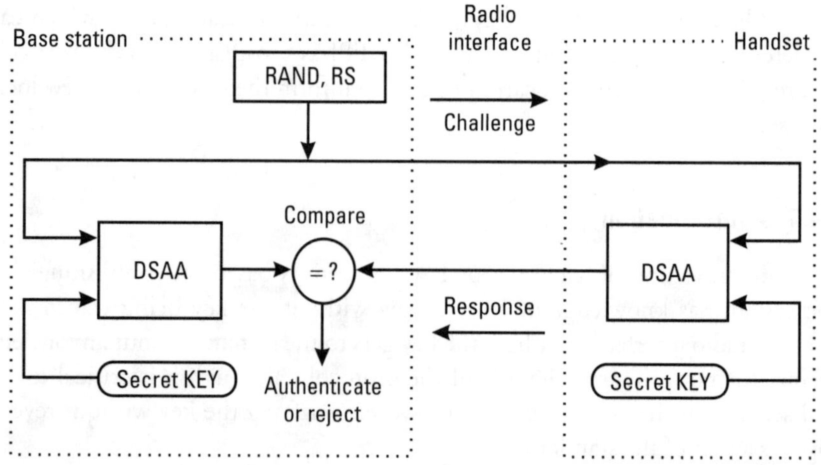

Figure 8.6 Handset authentication.

8.4.6 Encryption

The DECT system is designed to use the same algorithms it uses to perform authentication to derive session keys to perform encryption of the C-plane (signaling) and the U-plane (speech/data). The exact details of the process are given in ETS 300 175-7 and are not reproduced here.

8.5 Wireless ISDN

As presented to most users, ISDN appears as a digital signaling channel (a D-channel) of 16 kbps with two user channels (B-channels) of 64 kbps each. Wireless support for this 2B+D service is what we describe here.

There are actually two standardized ways in which the protocol provides interworking of ISDN equipment over the air interface, which are illustrated in Figure 8.7:

- End system configuration (see ETS 300 434 [6]), where the fixed part and the portable part together emulate a standard ISDN terminal;

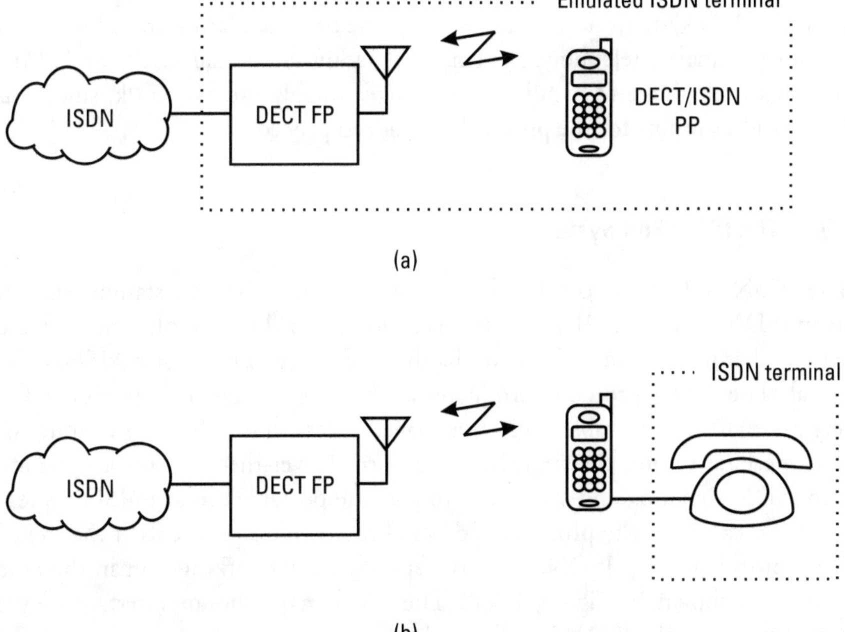

Figure 8.7 (a) DECT/ISDN end-system; (b) intermediate system.

- Intermediate system configuration (see ETS 300 822 [7]), where you can put a socket on the side of a cordless terminal into which you can plug a standard ISDN terminal.

In the first configuration, the ISDN protocol is terminated at the base station; in the second, the protocol passes through the base station and handset to be presented and terminated beyond the portable, very much like providing a radio local loop terminal with a remote ISDN socket.

The general method by which DECT carries ISDN is to define the following:

- How ISDN's two 64-kbps B-channels and its 16-kbps D-channel are mapped onto data packets at the air interface;
- How the cordless protocol stack interworks with the ISDN protocol stack;
- The means to keep the error rate in the B-channel low and comparable to that expected in a wired ISDN context.

DECT/ISDN systems need to use more of the base standard's procedures than do normal analog telephone systems. Upgrading a normal GAP- or CPAP-compliant telephone to handle ISDN is a reasonably extensive task, since you have to add support for the protocol's advanced procedures.

8.5.1 The ISDN End System

The ISDN end-system configuration is one in which the base station attaches to an ISDN basic rate (2B+D) interface, and the fixed system plus the portable act together to form an end system that has the behavior of a normal ISDN terminal. The ISDN protocol terminates at the base station. For its user traffic (e.g., speech), the portable is given managed access to the ISDN B-channel(s). It is the base station that translates the normal over-the-air control signaling into ISDN signaling and vice versa, to give the portable access to ISDN's features. To carry out the process of adapting the cordless protocols at the top of the protocol stack to ISDN, there is a specified interworking unit in the base station, as shown in Figure 8.8(a). The IWU maps the air interface's layer 3 messages onto the ISDN interface in both the send and the receive directions. There is already (by design) a reasonably good correspondence between DECT's services and those of ISDN, so the IWU's functions are not extensive.

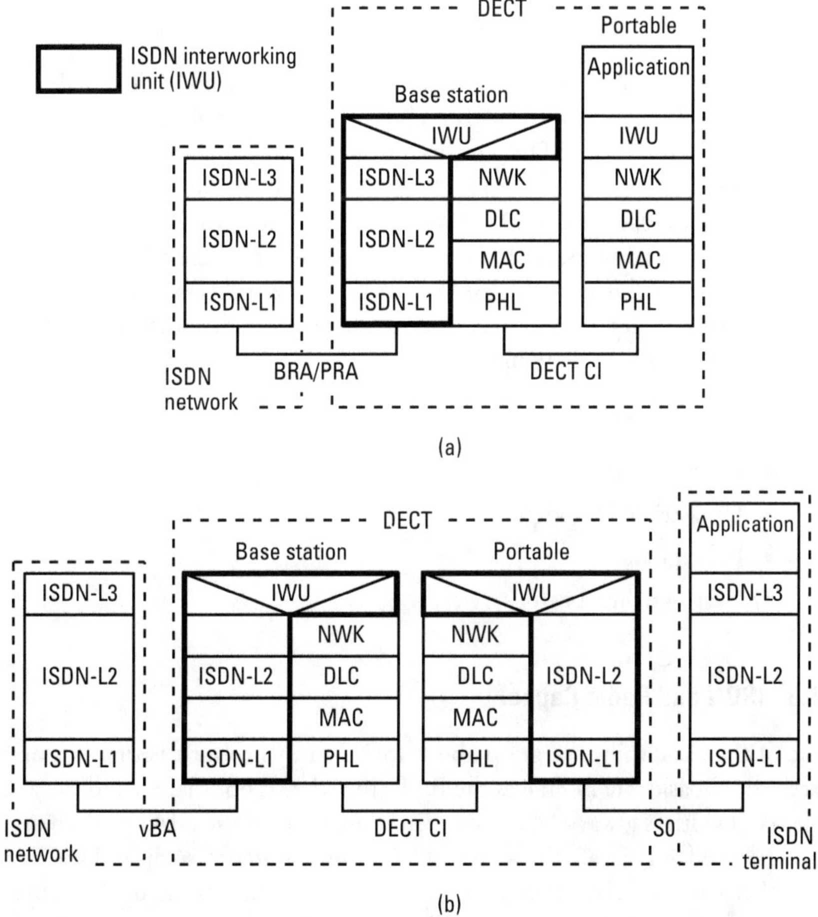

Figure 8.8 (a) DECT/ISDN end-system protocol stack; (b) intermediate system protocol stack. *Source:* ETSI [6,7].

In that configuration, the user's supported bearer services are as follows:

- Speech;
- 3.1-kHz audio;
- Unrestricted digital information at 64 kbps.

Applications using those bearer services may include normal 3.1-kHz telephony, 7-kHz telephony, group 4 fax, teletex, videotex, and voiceband data transmission such as group 3 fax and modem data transmission.

8.5.2 The ISDN Intermediate System

The ISDN intermediate system configuration is one in which the portable part provides a socket (an S-bus) into which an ISDN terminal can be plugged. It thereby gains access to ISDN and its features over the cordless air interface. Both public ISDN and private ISDN are supported. In this case, as shown in Figure 8.8(b), a specified ISDN IWU exists within both PP and FP, which maps ISDN layer 3 messages bidirectionally to and from the air interface. The IWUs are specified to give the ISDN user transparent access to ISDN services and functions. A larger list of bearer services is supported, compared to the end system, including the following:

- Speech;
- 3.1-kHz audio;
- Unrestricted 64-kbps data;
- Packet data;
- User signaling bearer service.

8.5.3 ISDN and Radio Capacity

One of the basic things to grasp about ISDN in this context is that in normal fixed ISDN basic rate interfaces the PHL (the wire) is not shared, and the available bandwidth is always 144 kbps even if only one of the 64-kbps B-channels is to be used for a call. In a shared radio environment such as DECT (or any other wireless access technology), it is not acceptable to tie up 144 kbps of capacity for two B-channels and a D-channel if only one B-channel is to be used. The procedure for setting up a normal call on ISDN is to establish layers 1 and 2 of the ISDN protocol before the layer 3 protocol message may be sent asking for the specific service needed. That is the first point in the call-setup procedure, that you get an idea of how much bandwidth is really needed. For DECT therefore, it is the ISDN layer 3 message that determines the amount of radio spectrum to be used for the call.

When the bearer service requested is 64 kbps unrestricted digital information, the DECT or PWT system sets up just that, a 64-kbps channel based on the LU7 service at the DLC. The LU7 service uses buffering and retransmission and occupies more than 64 kbps on the air interface to be able to provide some error correction to compensate for radio propagation problems. Since the user's bearer service is labeled "unrestricted," we have to give it a full 64-kbps radio bearer of the best possible quality. However, in the case that the user's bearer service is for speech, it is permitted to save radio bandwidth by transcoding the

ISDN 64-kbps speech channel into 32 kbps using the standard ADPCM encoding. In that case, no extra error protection is provided over and above that normally provided by DECT.

8.5.4 ISDN B-Channel Carried Over the LU7 Service

The requirements for error performance in ISDN services are specified by the ITU-T in Recommendation G.821 [11]. The reader will find that the error requirements are difficult to achieve with a radio-based PHL. While the ISDN signaling (the ISDN D-channel) has its own error protection and tolerates errors even if the error control of the control-traffic plane of the cordless protocol stack were to fail, the ISDN user traffic in the B-channel has no specific error protection, and neither does the regular user-traffic plane of the protocol stack when configured by voice traffic. For that reason, a DLC-layer service is defined (see Section 6.2.3) as imporoving the error rate of the user-traffic plane of the protocol stack. It does that by using forward error correction (FEC), with an ARQ retransmission scheme for those frames of user data that still cannot be corrected. This service is called the LU7 service (see Table 6.3).

To be able to retransmit some parts of the 64-kbps B-channel if they have too many errors, the gross data rate on the air interface needs to be greater than 64 kbps. Hence, LU7 is carried over a P80 physical packet, that is, a double slot on the air interface, having 64 bits of protected signaling and 800 bits of user data (see Figure 4.5). Within those 800 bits, 720 bits (i.e., 72 kbps) ultimately are available for carrying the 64-kbps ISDN B-channel data. The 64-kbps data are buffered 80 ms at a time, sent over the LU7 service at 72 kbps and in the time remaining out of the 80 ms, the recipient asks for a retransmission of any data it cannot correct using the FEC. If there is not too much data to retransmit, the recipient will finally have all the data in its 80-ms buffer in time to send them out fully corrected. This method of keeping the error rate low introduces an 80-ms delay into the user traffic.

Technically, at the MAC layer, the unprotected multiplex U80a is used in a symmetrical single-bearer connection. That gives a raw data rate of 80 kbps to be used for the 64-kbps channel. The advanced MAC connection control and B-field signaling are used, and paging uses the full format (advanced features not covered in Chapter 5). At the DLC layer, the FU7 frame is used (Figure 8.9).

8.6 Data Applications

DECT and PWT have a wide range of wireless data capabilities that can be described here only in outline. A family of data application profiles ensures

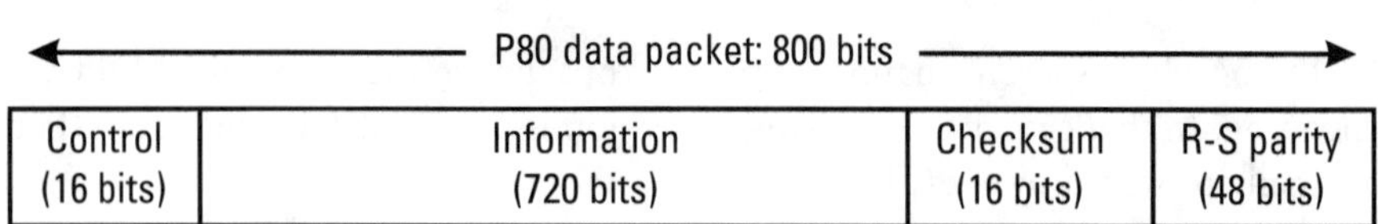

Figure 8.9 Mapping ISDN 64-kbps service into the FU7 frame.

interoperability between products from different manufacturers for a wide range of data systems. The data applications all exploit the advanced services of the base standards that are not needed for speech but that are there for purposes such as this. Those advanced services are specifically oriented toward LAN, multimedia, and serial data capability. Each member of the data application profile family has been optimized for a different kind of user service, and distinction is made between services without mobility (called class 1) and those with mobility (called class 2). The different data profiles are modular and closely related, so they can be implemented economically and efficiently either independently or in combination. The services and relationships of the different profiles are fully described in ETR 185 [12], and some of the capabilities are illustrated in Figure 8.10.

The data profiles provide security (authentication and encryption), call charging, flexible throughput up to 552 kbps, high reliability, error correction, and other features, thus making them suitable for public wireless data services as well as data services in the private and business environments. Although it would be impossible to do justice to the entire range of data applications here, the family of applications comprises the members briefly described next.

8.6.1 Generic Frame Relay Services

The generic frame relay services, where frames of data are transferred with their boundaries preserved but without any notification of receipt, are the basis for all the connectionless data services defined by the protocol. The basic standard for the interworking of nonmobile application to connectionless networks within closed user groups is the ETS 300 435 data services profile [14], which defines service types A and B, class 1. It defines a basic frame relay service and then includes annexes for interworking with Ethernet and token ring LANs at a throughput of up to 552 kbps. The distinction it makes between type A and type B services is based on maximum throughput. The type A service is simpler and uses fewer of the base standard's features but has a maximum throughput of 24 kbps. The type B service uses more of the protocol's optional lower-layer

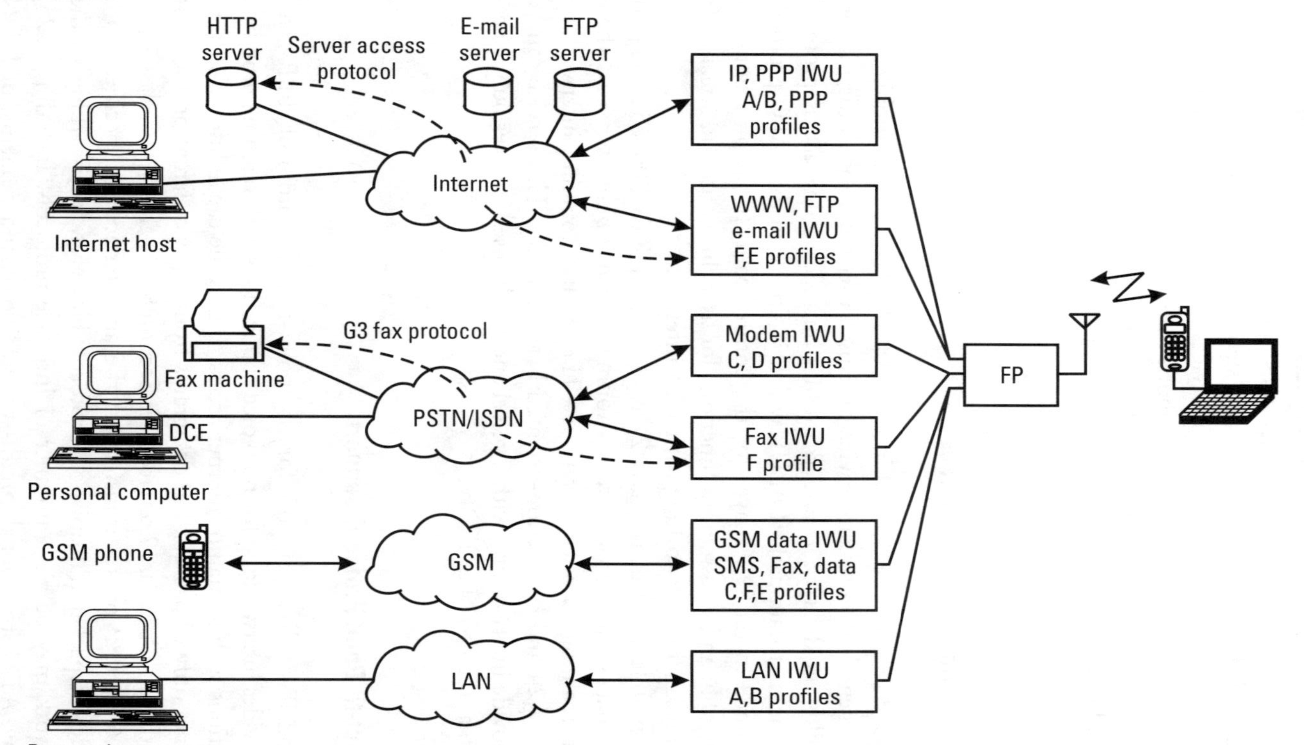

Figure 8.10 Data capabilities. *Source:* ETSI [13].

capabilities but has a maximum throughput of 552 kbps. In providing the higher throughput rates, it uses larger packets on the air interface and more of them, so equipment using type B service consumes more power.

For generic frame relay services where significant mobility is required, the basic standard is the ETS 300 701 data services profile (service types A and B, class 2) [15]. The application profile supports similar services to the class 1 but is not restricted to closed user groups. As a result, in addition to its support for Ethernet and token ring LANs, direct interworking with Internet protocol (IP) networks is defined.

8.6.2 Generic Data Stream Services

The type C, class 1 service defined in ETS 300 699 [16] is a nonmobile, nontransparent generic data stream service for closed user groups with high integrity. It builds on the type A and B profiles but adds a link access protocol service in the user data path (LAPU), which is similar to the LAPC protocol already defined in the DLC layer for the control path. It has provisions for packet assembly and disassembly functions for asynchronous data streams. It includes annexes for interworking with V.24 interfaces.

The mobile equivalent (class 2) is defined in ETS 300 651 [17]. It extends the data stream service into environments such as public services, where significant roaming is a characteristic. The profile contains interworking annexes to V.24 and connection-oriented bearer services. This service can be used to provide interworking with a voiceband modem service over public networks such as the PSTN or ISDN.

8.6.3 Point-to-Point Protocol Support

DECT and PWT have support for the point-to-point protocol, PPP [18], which provides interworking for nonvoice applications with roaming mobility, in applications such as dialup Internet access. It uses the capabilities of PPP to transport not only IP packets, but datagrams from many different protocols. It builds on the type A/B class 2 services (see Section 8.6.1) and type C class 2 (see Section 8.6.2). PPP packet transfers on the air interface are specified via a highly efficient packet transmission protocol. However, interworking to the fixed network may be via a number of interface protocols, including X.25, frame relay, ATM, traditional circuit-switched voiceband modems, or ISDN connections.

8.6.4 Mobile Multimedia Support

The ETS 300 755 data services profile [19] is a multimedia mobile messaging service with specific provision for facsimile services (service type F, class 2). It creates high-level interoperability for a range of services, including fax, e-mail, WWW, HTTP, and FTP. It does that through a multimedia file transfer mechanism built on the generic data stream service (see Section 8.6.2), with full support for roaming and public access. One of the main applications of this profile is to provide interworking to public and private group 3 fax services.

8.6.5 Low-Rate Messaging

The low-rate mobile messaging service (service type E, class 2) provides a means for the low-rate and low-power-consumption transfer of different types of messages, including alphanumeric paging messages and the transfer of multimedia message objects. The ETS 300 757 data services profile [20] provides both point-to-point and point-to-multipoint messaging through the signaling channel, with and without acknowledgment, based on the multimedia messaging service (MMS) specified in ETS 300 755 [19]. Because it uses only the control plane, it can easily coexist with applications delivering voice service, such as a GAP handset. No user-plane functions are required, so there is no interruption of the voice service. This service can be used for applications that provide private and public roaming message services, such as the GSM SMS [21].

8.6.6 Isochronous Data Bearer Support

The D2 profile [22] is suitable for transparent, isochronous transfer of synchronous data streams and is intended for use in private and public roaming applications. Different qualities of service are specified from unprotected to fully protected, providing different levels of error performance and different levels of complexity. Interworking to isochronous modems and standard synchronous circuits is the aim of this profile as well as video telephony, video conferencing, and end-to-end encrypted telephone services over external networks. The D1 profile [23] provides a nonmobile equivalent service to the D2 profile for closed user groups.

8.7 Radio Local Loop Applications

RLL generally refers to the provision of a telephony service to a "standard telephone" by use of a radio interface to extend the reach of the network providing service (Figure 8.11). The need for copper wire in the final part of the connection from the local exchange is removed, and an expensive part of the access network is eliminated. Of course, the radio installation itself can be expensive. However, it usually is assumed that, over the long term, the installation and maintenance costs of an RLL installation will work out cheaper than copper wires. In this scenario, all the cordless equipment is part of the network itself, and the network operator's responsibility normally ends at the telephone socket on the cordless terminal adapter (CTA), which is, therefore, the network termination point (NTP).

ETSI's report on DECT RLL (ETR 308 [24]) examines in detail the specific services that may be offered. It identifies the basic wired analog PSTN services that could be replaced by an RLL system and also identifies that there are market opportunities for much more advanced services and different service scenarios. For example, it is possible for the CTA to belong to the customer rather than to the network operator. In this case, the NTP is in the air interface between the fixed part and the CTA. The operator's responsibility for the system, therefore, ends with the provision of a suitable radio signal strength at some point on the customer's premises, and the customer takes responsibility for the CTA's siting, maintenance, and performance. This scenario has lower installation and maintenance costs for the operator, but with inexpert customers the likelihood is that ongoing service quality issues may arise because of bad CTA installation.

The basic RLL applications (PSTN replacement including modem applications up to 33.6 kbps) are covered in part 1 of the RLL Access Profile (RAP) [25]. More advanced applications of RLL (e.g., when the attached equipment is an ISDN terminal or an ISDN PABX) are covered in part 2 of the RAP [26], which also covers requirements for broadband packet data applications up to 552 kbps.

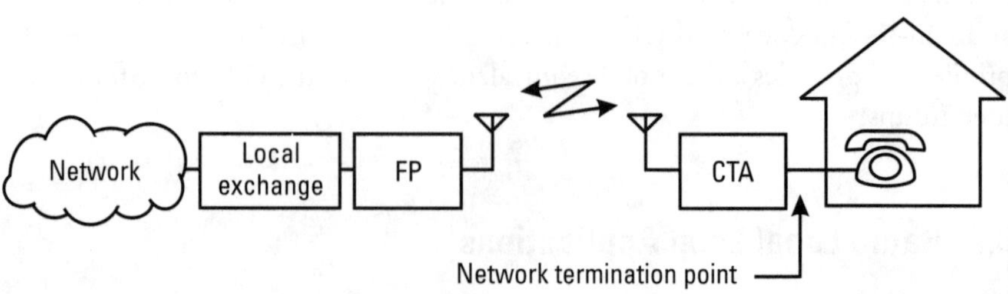

Figure 8.11 Basic RLL service delivery scenario.

8.7.1 Basic RLL Applications

In general, the "portable part" of an RLL system, the CTA, is not really portable in the sense of a normal cordless portable part. Usually, it stays in a fixed location. Also unlike most normal terminals, a CTA could provide multiple remote analog lines, suitable for interfacing to a PABX. Basic RLL service requires a CTA that is closely based on GAP, with minimal changes and additions. The basic changes are as follows:

- Some user-originated signaling information (e.g., pulse dialed digits and register recall) and some local exchange–originated signaling (e.g., meter pulses and line reversals) need to be specially transferred across the air interface since the ADPCM-encoded channel cannot carry it.
- GAP features not relevant to RLL (e.g., partial release) need to be removed.
- Call clearing is modified to meet the requirements for the emergency services to take control of clear-down in emergency calls.
- Support is added for the automatic invocation of a 64-kbps bearer service to enable the use of fax and modems (up to 33.6 kbps), since 32-kbps ADPCM is not transparent to modem protocols above certain limits.
- Features to allow for operations administration and maintenance (OA&M) are added.

Note that, although it is based on GAP, an RLL CTA will not interwork with a purely GAP fixed part. The fixed part needs to supports the GAP and the RAP extensions.

8.7.2 Advanced RLL Applications

For advanced RLL services, part 2 of the RAP simply refers to existing application profiles for the optional provision of services, including:

- The DECT-ISDN intermediate system (see Section 8.5.2) for offering an ISDN basic rate service (the ISDN intermediate system standardization work will later include interworking of ISDN primary rate access, suitable for interfacing to ISDN PABXs).
- The DECT data profiles (see Section 8.6) for providing Internet access, modem support; and group 3 fax support.

For speech service, RLL applications have the same spectrum efficiency as all other cordless services using 32-kbps ADPCM. For nonspeech applications, advanced RAP services provide direct efficient transfer of data without the need to digitize modem signals. That is much more spectrum efficient than, for example, modem signals over 32-kbps ADPCM. For packet-oriented applications, the data profiles allow for the air interface resources to be released when there are no data to send, providing even better use of the spectrum.

In addition, features have been introduced into the advanced RAP application profile for the OA&M of CTAs supporting these services. The OA&M features are largely an extension of those defined in the basic RLL services profile.

8.7.3 Service Delivery Scenarios

The basic service scenario in Figure 8.11 (wherever the NTP lies) is by no means the only possible way to use DECT or PWT to deliver fixed services. Two more are shown in Figure 8.12, and many more are possible (see ETR 308 [24]). In Figure 8.12(a), the radio repeater, the wireless relay station (WRS) might be used to extend the range of the basic RLL system.

The cost of an RLL system strongly depends on the cost of base stations, their sites, and the wiring of the sites. In covering difficult subscribers (from a radio propagation point of view), it may be easier to use a repeater (where a network connection to the repeater site does not have to be provided) than to install a fully networked base station.

In the second situation, illustrated in Figure 8.12(b), a WRS might be built into the CTA, to offer customers cordless telephony within their premises, without the customers having to install a separate cordless telephone. In principle, the extra facilities needed to add wireless relay functions to a CTA are relatively few, and the cost of doing so is likely to be economic compared to having a separate CTA and cordless base station.

8.8 Wireless Relay Stations

The WRS is a special unit capable of relaying DECT radio transmissions. The WRS works by linking two radio connections working on two different timeslots. Therefore, it can extend base station coverage without requiring a network connection to a base station controller. A WRS utilizes the normal way DECT accesses the RF spectrum. The full dynamic channel allocation functionality is available to each of the links, and nominally, information is transparently relayed through the WRS.

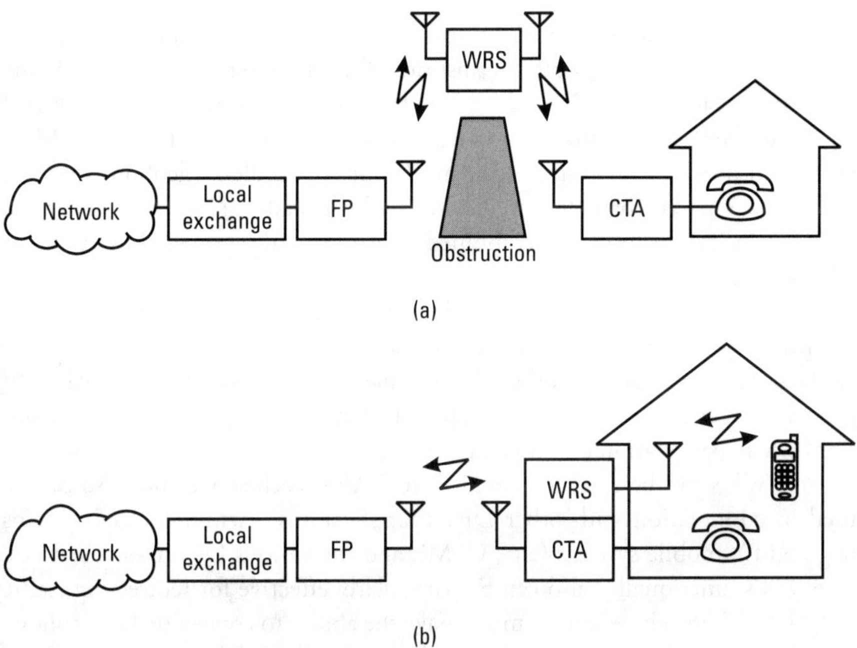

Figure 8.12 (a) RLL delivery via a WRS; (b) combined WRS/CTA.

There actually are two ways of implementing WRSs using the DECT protocol: the "repeater part" (REP) and the "cordless radio fixed part" (CRFP). The ETSI technical report on repeaters (ETR 246) [27] and the WRS standard itself [28] provide details of the specific difference between the two types. In principle, the WRS acts toward a PP exactly as if it were an ordinary RFP, so that a PP cannot distinguish between a WRS and an RFP. Similarly, the WRS acts toward an RFP just like a portable. This principle can be used to put in place a single repeater between an FP and a distant PP or even several repeaters where the distance between PP and FP is very large.

8.8.1 The Basic Uses of a Repeater

The WRS is suited to providing cost-effective infrastructures. Often, it is better than using a standard base station in low-traffic-density applications, where it is used to extend or improve coverage indoors or outdoors that otherwise is marginal or to extend coverage to areas behind obstructions.

A typical application, illustrated in Figure 8.12(a), is the WRS extending RLL coverage to an otherwise difficult-to-cover area. In an RLL application, normally, a cordless terminal adapter placed at roof level might feed a standard copper pair into the residence for users to connect one or more wired telephones. In another application, instead of feeding service into the home via copper cables, a WRS used at rooftop height instead of the standard CTA can feed service into the home as a cordless radio link (Figure 8.12(b)), leading to much less wiring and the availability of cordless telephony directly, without a separate base station having to be installed. The two functions, WRS and CTA, even can be combined to economically provide both types of RLL access.

In larger residences, where traffic is normally low compared to commercial premises, it may be much easier to install a repeater to extend radio coverage rather than install what otherwise may have to be a small two-cell PABX. In general, it is cost effective to use a repeater (versus putting in another base station) where base station density and traffic are low.

A WRS can be used to extend basic GAP speech service but also can be used in conjunction with other DECT applications, including fixed systems (e.g., RLL), mobile systems (e.g., GSM), and the various DECT data services. The WRS functionality also can be particularly effective for redirecting traffic in PABXs where a base station might have the ability to cope with 12 telephone calls, but the backhaul link to the cordless controller might offer, say, only 8. The cell's extra capacity could be used to farm out excess local traffic to another cell for backhaul to the switch.

8.8.2 Radio Spectrum Implications

The repeater uses more radio spectrum for one link compared to the use of a base station sited at the same point (Figure 8.13). This is clear if you think of a simple duplex voice conversation. Normally, it would use just two physical channels, one for the uplink and one for the downlink. However, for a repeater, four channels are used.

Of course, a repeater with a simple omnidirectional antenna provides a relatively small increase in the overall coverage area of a system, since much of the repeated cell area overlaps the original cell area. However, a repeater with an antenna having a single high-gain lobe as well as an omnidirectional pattern in other directions extends the area further. Naturally, both uses add the same extra load onto the spectrum. The extra channels are now in use over much the same area, regardless of the shape of the area. As a result, some national regulatory authorities limit the use of the repeater to just one extra hop. That limit is also set forward in TBR 10 [29], but national authorities may extend it.

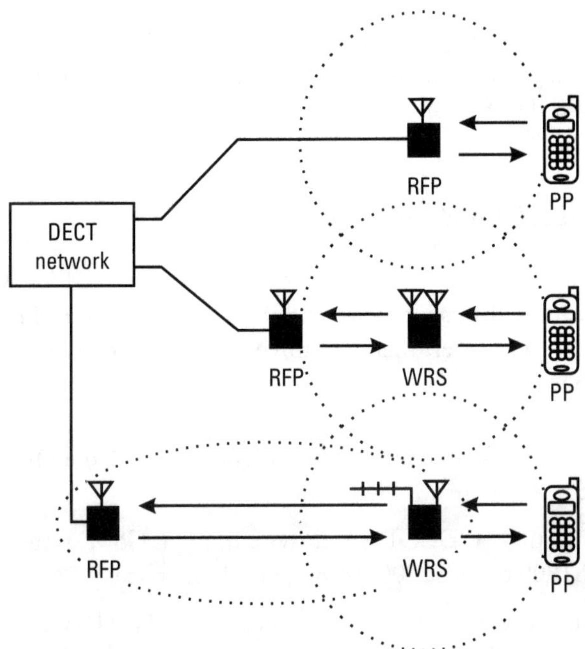

Figure 8.13 Additional radio spectrum used by a WRS.

8.8.3 Transmission Delay

Because the WRS does not retransmit in the same timeslot as it receives, a small transmission delay is always introduced. In case of multiple WRSs working in series, the delays sometimes can add up to unacceptable lengths. The two WRS types differ in the amount of delay they introduce. The addition of each CRFP into a normal B-field connection adds 5 ms in each direction. The REP adds less delay: a 2.5-ms one-way delay for any number of REPs in cascade.

Excessive delay can have serious effects on how users perceive speech quality, primarily due to electrical or acoustic echoes from certain parts of a complete end-to-end connection, notably the terminations of two-wire sections. Because of the way the REP-type WRS works, any number of relays in series is tolerable. However, even with the CRFP form of the WRS, one radio relay is considered tolerable within the standard DECT echo-control system; for two or more in series, the essential additional echo-control mechanisms are specified [5].

As for data connections, delay sometimes can be critical. In many cases, however, a few extra milliseconds make little difference. Because of the many types of data connection and the plethora of applications, it is not possible to generalize in the same way as is possible about speech. An individual assessment is needed about the effect of relays on each application.

8.9 DECT/GSM Interworking

The treatment here of such a complex application as DECT/GSM interworking cannot be more than superficial, and there is a full ETSI technical report on the subject [30]. However, there are three basic ways to combine DECT systems with GSM:

- Using a dual-mode DECT/GSM handset where the DECT and GSM fixed systems are not interconnected;
- Connecting a DECT fixed system to a GSM system (GSM900, DCS1800 or PCS1900) through its A-interface;
- Connecting a DECT fixed system to a GSM system through an enhanced ISDN interface.

There are many other enhancements to the above scenarios, two of which we cover here:

- Using a dual-mode DECT/GSM handset where the DECT and GSM fixed systems are also interconnected;
- Using the mobility features of the GSM fixed system to support DECT-only networks.

8.9.1 Dual-Mode DECT/GSM Terminal With Independent Fixed Systems

One of the simplest ways to take advantage of a combination of DECT and GSM is simply to have a hybrid DECT/GSM handset (Figure 8.14). In that scenario, there is no direct connection between the GSM cellular network and the DECT cordless system. A user would have a normal GAP-compliant DECT cordless handset that is physically integrated with a GSM terminal. The portable parts would share case, display, keypad, battery, and, in the case of a DCS1800/DECT dual-mode terminal, perhaps much of the radio subsystem. The terminal has two subscriptions: one in a GSM subscriber identity module (SIM) and one in a DECT authentication module (DAM). The two subscriptions might be combined in one smart card.

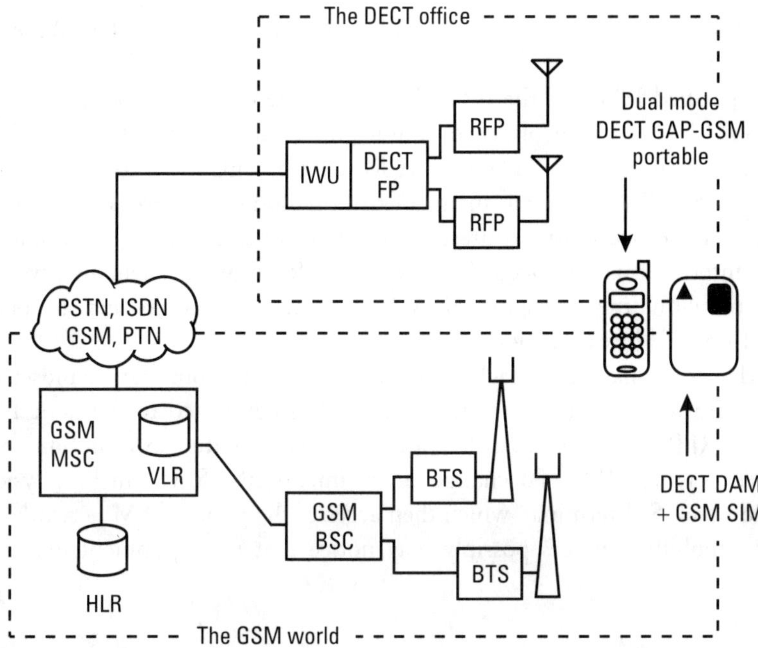

Figure 8.14 DECT/GSM dual-mode terminal without network interconnection. *Source:* ETSI [30].

In the first instance, there need not be any coordination between the DECT and GSM parts of the terminal. The user manually selects which of the systems to use for outgoing calls and may, for example, manually set up call forwarding within the GSM and DECT fixed systems to deliver calls from one system to the other whenever there is no answer (being careful to avoid forwarding loops).

In a more complex scenario, the two applications in the same handset could communicate with each other. The handset would sense when it is in the coverage area of each system and would then automatically select which system to use for outgoing calls. When both systems are available, the handset would select whichever is either explicitly preferred by the user, or perhaps just the cheaper to use. For incoming calls, the handset would automatically set up the call forwarding service for each system, depending on its known environment.

8.9.2 DECT System Connection to GSM's A-Interface

When it comes to integrating DECT and GSM fixed infrastructure, the first way is to attach an FP to the GSM MSC at one of its standard A-interfaces

(Figure 8.15). That is where the GSM base station controllers (BSCs) are normally attached.

Specified IWUs in the FP and PP make the translation between DECT and GSM layer 3 protocols, according to the requirements of the GSM interworking profile (GIP). To the GSM MSC, the DECT FP looks just like a normal GSM BSC. One advantage of this method of interworking is that no changes are needed in the A-interface or in the MSCs. Nevertheless, changes to the A-interface may be needed in the future for more efficient interworking (e.g., to allow the transport of ADPCM-encoded speech and to permit DECT-specific bearer services).

The terminal that attaches to the FP is a GIP-compliant handset that accepts a SIM card. It can roam between RFPs connected to the same FP or between RFPs connected to different FPs by involving the GSM mobility management in the MSC. To change environments, the SIM can be physically moved to a GSM terminal, which then accesses the public GSM network. The GSM's mobility services possibly may notice that the equipment identity is

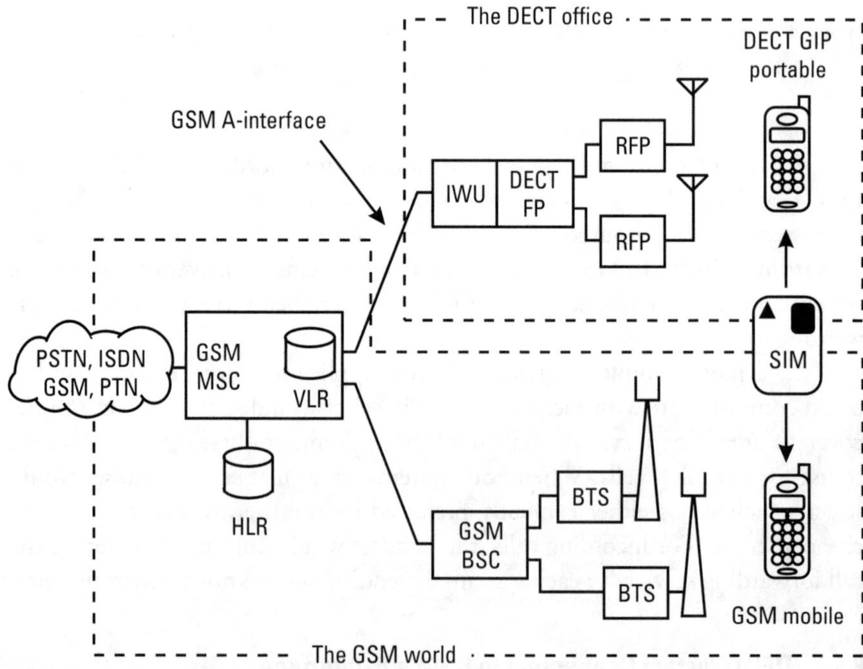

Figure 8.15 DECT/GSM fixed-system interworking at the GSM A-interface. *Source:* ETSI [30].

different, but service that follows the user identity in the SIM card is completely transferable between the DECT and the GSM handsets.

8.9.3 DECT System Connection to GSM Via an ISDN Interface

The second method of integrating DECT and GSM infrastructure is to make the connection between the DECT system and the GSM network via ISDN (Figure 8.16). This method of interworking is particularly suitable for connecting DECT PABXs, which normally support a DSS1 network interface but not the GSM A-interface.

The original Digital Subscriber Signaling System No. 1 (DSS1) interface alone does not support authentication and mobility management, but ETSI has defined those enhancements, along with the support for other more general cordless terminal mobility functions to create the DSS1+ interface [8].

An ISDN interface between GSM network and DECT system allows GSM service providers to offer their services via existing DECT access networks that have ISDN interfaces. The DECT access network could be owned by the GSM operator and leased to the customer, or it could be a customer-

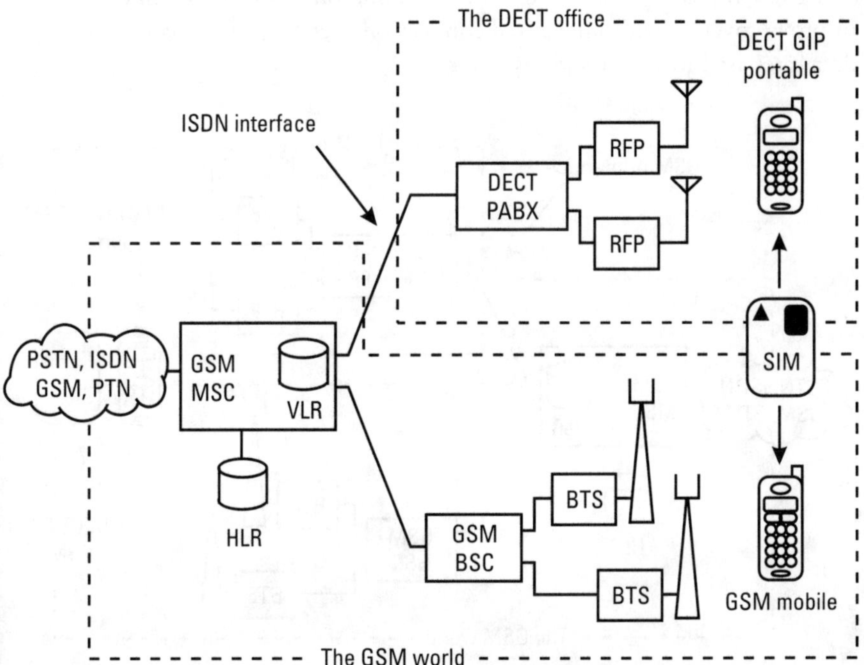

Figure 8.16 DECT/GSM fixed system interworking via ISDN. *Source:* ETSI [30].

owned PABX or private network, which can be upgraded to provide access to GSM. By swapping the SIM between terminals, the GSM user can access telecommunication services in any location with DECT coverage where access is allowed. Local mobility is managed by the DECT system and wide area mobility by the GSM network.

8.9.4 DECT/GSM Dual-Mode Terminal With Interconnected Fixed Systems

We have seen how DECT and GSM systems may be interconnected via either the GSM A-interface or an enhanced ISDN interface and also how service can be obtained by swapping SIM cards between separate DECT and GSM terminals. It is, of course, also possible to use dual-mode terminals with interconnected systems (Figure 8.17). The SIM swapping between the DECT and GSM parts of the handset is, in effect, done automatically, which is much more convenient for the user.

Such a system may be useful where the GSM operator needs to add capacity in high-density areas (such as a wireless office) or where higher data rate services are needed and not provided by GSM. The capacity of DECT can be as high as 10,000 Erlang/km^2/floor. DECT RFPs also are expected to continue being much less expensive than GSM base stations. WRSs also may be used to improve coverage without a wired connection. Better speech quality and higher data rates are further attractions.

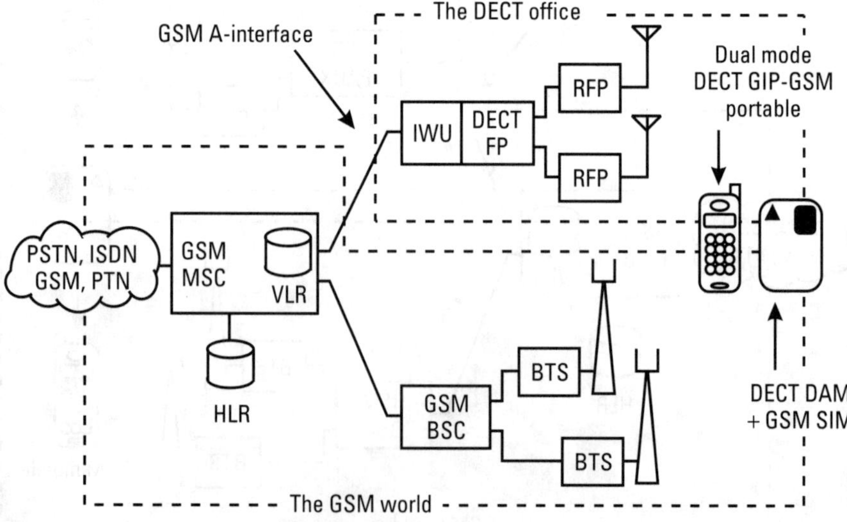

Figure 8.17 DECT/GSM fixed system internetworking with a dual-mode terminal. *Source:* ETSI [30].

In this basic scenario, the DECT FP is interconnected to the MSC by the A-interface. The user has a GSM subscription and a dual-mode terminal and can roam between FPs and BTSs. Handover (involving the MSC) is possible between RFPs connected to different FPs, but there currently is no standard for handover between an RFP and a BTS. The DECT part of the dual-mode terminal is a GIM-compliant portable, and both parts use the same subscription (a single SIM).

The same DECT installation also can be used for RLL access to the GSM network using cordless terminal adapters complying to the RAP.

8.9.5 DECT-to-DECT Connection Via GSM

With the DECT/GSM interworking profiles, it is even possible to provide mobility services for just DECT systems (Figure 8.18). In that scenario, the DECT PABXs are connected to a GSM MSC using the DSS1+ ISDN interface. Traffic is carried by the GSM network, and inter-PABX mobility is provided by the GSM HLR/VLR. The terminal is a GIP-compliant DECT portable.

In an equivalent scenario, we have the same functions as in Figure 8.18, but a DECT FP is connected to the MSC by the A-interface. Local mobility is provided by the DECT system. Wide-area mobility is provided by the GSM HLR/VLR, and the terminal is again a GIP-compliant portable. This scenario also can be extended with a dual-mode terminal, so that when "on site" the call is made via a DECT FP (the wireless PABX), and when "off site" via GSM. Such a service might use existing public pedestrian service sites (e.g., in

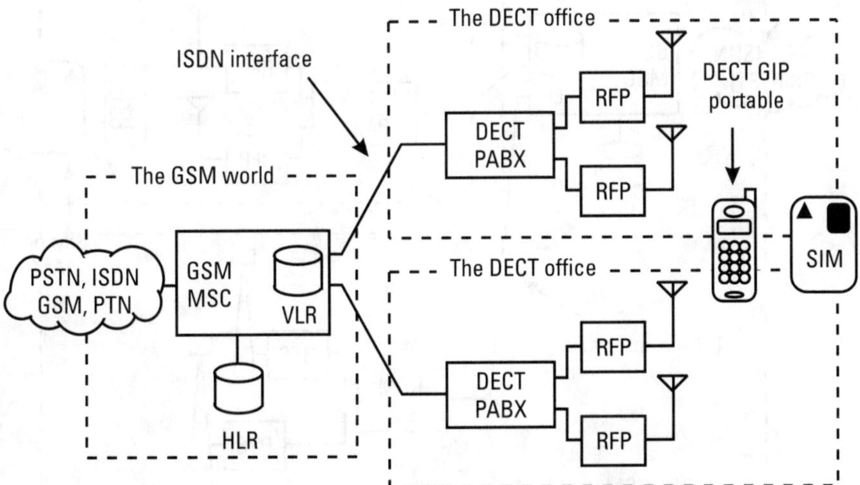

Figure 8.18 DECT/DECT private networking via GSM. *Source:* ETSI [30].

shopping centers and metropolitan train stations), but subscribers would have incoming call capability and mobility.

In a more advanced scenario (Figure 8.19), a roaming agreement exists between a GSM operator and DECT system owners. The DECT system may be a single FP/RFP (e.g., a residential system) or a wireless PABX with multiple RFPs (e.g., an office environment). The user has a DECT or GSM subscription with either party and can roam seamlessly between the GSM network and the DECT system. The terminal is a dual-mode GSM mobile and GIP-compliant DECT portable. A SIM or a DAM is used, depending on whether the subscription is with a DECT or GSM operator.

8.10 Public and Public/Private Applications

The DECT/PWT system as a whole does not really distinguish between public and private applications. However, a number of things are specified to

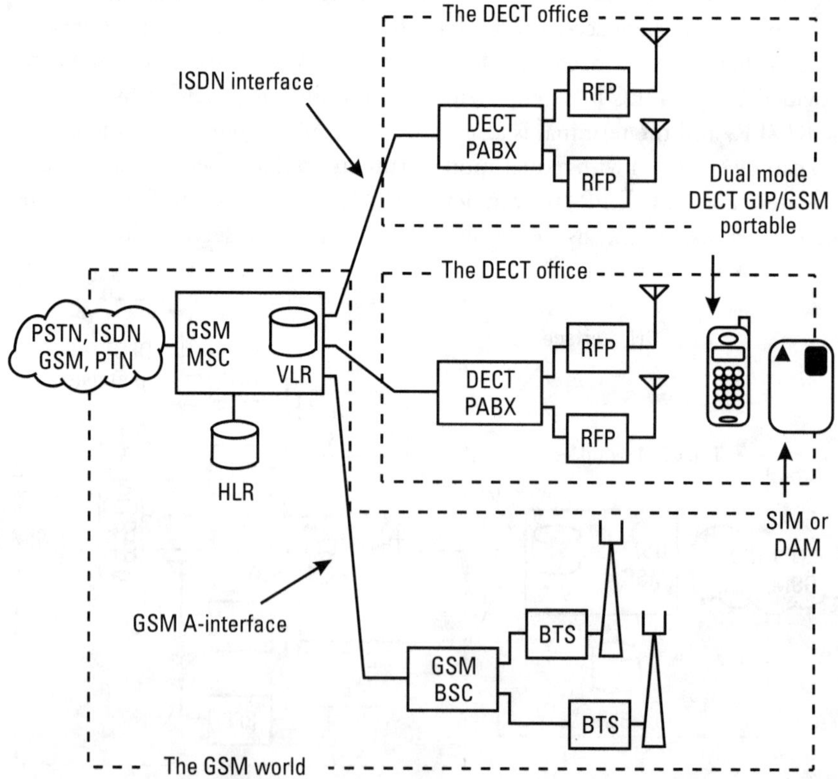

Figure 8.19 DECT/DECT public/private interworking via GSM. *Source:* ETSI [30].

make public systems viable and the integration of public and private systems possible.

8.10.1 Public Access Systems

Providing access to a public cordless infrastructure is much the same, technically at least, as providing access to a private system. The concept, sometimes called telepoint, is that of providing, in effect, a universal public telephone booth, for which the users all have their own handsets. That sounds attractive if handset ownership gets to be common, since the operator no longer has to maintain equipment that is accessible to the public and subject to misuse and other related problems.

The commercial problem that has been seen in the past with such services is that they have never been deployed with continuous coverage because of the cost of acquiring and equipping many small cells. Hence, the ubiquity and falling cost to subscribers of cellular radio access means that not many public telepoint systems are still in operation.

The critical technical matter for public access is that of authentication for the purposes of billing. The procedures built into the air interface for challenge-response authentication (and encryption for privacy) take care of that. However, what the standards do not cover is the key-handling and key-distribution system. As with all secure systems, the technically exciting bit in the center that does the authentication or encryption is not the most critical part.

It used to be thought that authentication and encryption were issues for public applications only, but increasingly the security of private PABXs is being compromised by persistent and technically adept hackers. Thus, the security features of public systems, already built into the DECT system, are being increasingly applied to private applications.

8.10.2 Integration of Public and Private System Access

A possibly more attractive use of public access cordless systems is to allow seamless roaming between public and private infrastructure, called cordless terminal mobility or CTM in the European standards arena.

The purpose of CTM is to enable users of cordless terminals to make and receive calls from any location within any public or private telecommunication network where coverage is provided. It is intended that it should be possible to add CTM functionality to those parts of a private telecommunications network where it is required without modification to any other part of the network.

Mobility functions are confined to two places for a given cordless terminal (Figure 8.20):

- The home area, which comprises the HLR within the cordless exchange that controls the home private telecommunication network;
- The currently visited area, which comprises the VLR within the exchange that controls the visited network and the part of the radio system in the vicinity of the portable.

In providing cordless terminal mobility, it was a goal that no changes should be necessary in any part of the private telecommunications network not involved in providing mobility.

A roaming cordless terminal is associated temporarily with an access identity at the network it is currently visiting, which persists while it is registered there. Knowledge of the visited area is sent to the cordless terminal's HLR at the time of its first location registration at the visited network. The access identity to be used is stored in the VLR in the visited area and is updated locally when the access details change (such as roaming within the visited network).

Each cordless terminal is permanently associated with its home network as if it were a terminal attached to that network. Thus, the cordless terminal can be made an addressable entity by assigning a telephone number to it from the

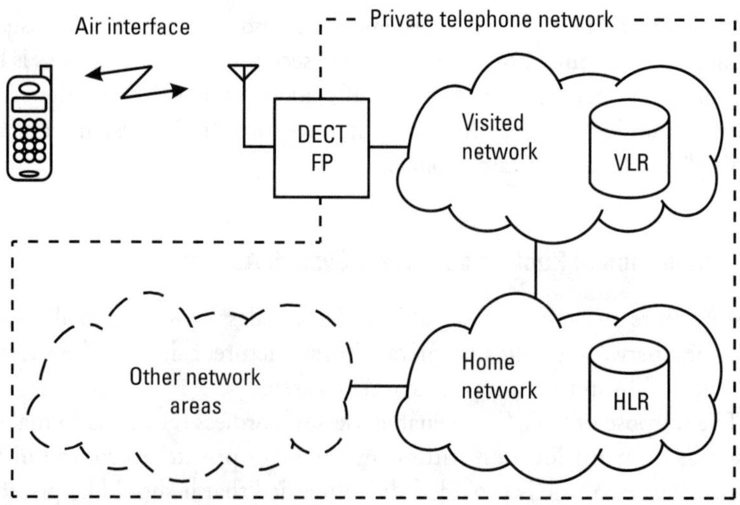

Figure 8.20 The system model for providing CTM.

number range at the home network, which allows every other connected network to route calls to that cordless terminal.

To accommodate certain implementation scenarios, it is possible to assign to a terminal an identifier that is permanent but not a telephone number on its home network. Such a facility would be used with cordless terminals that are unable to support user-assigned identifiers or in situations where interworking with other (public) networks is a requirement. The identifier is used to determine the CTM user's entry in the HLR and has a one-to-one relationship with the telephone number that is used for routing purposes.

Note that this does not prevent scenarios where the same number can be assigned to a user's cordless terminal and fixed terminal at the same time. The home network can take a local decision which of the two is to receive an incoming call or it can decide to ring at both terminals in parallel.

The DECT standard that specifies the set of technical requirements for FPs and PPs necessary for the support of CTM is the CAP [31], which is an extension of the GAP, specifying only those components not mandatory in the GAP that are needed in the CTM context. It supports telephony teleservice and provides a 32-kbps ADPCM speech bearer service. The concepts of the CAP are similar to those covered by the use of the GSM network as the provider of mobility services for multiple public and private cordless systems (see Section 8.9.5 and Figure 8.20).

References

1. ETSI, *Digital Enhanced Cordless Telecommunications (DECT); Common Interface (CI),* ETS 300 175 (9 parts), Sophia Antipolis, France, September 1996.

2. TIA/EIA, *Personal Wireless Telecommunications Interoperability Standard (PWT),* 13 parts, TIA/EIA 662-1997.

3. ETSI, *Digital Enhanced Cordless Telecommunications (DECT); Generic Access Profile (GAP),* EN 300 444, Sophia Antipolis, France, August 1997.

4. TIA/EIA, *Personal Wireless Telecommunications Interoperability Standard (PWT); Part 9: Customer Premises Access Profile (CPAP),* TIA/EIA 662-9-1997.

5. ETSI, *Digital Enhanced Cordless Telecommunications (DECT); Common Interface (CI); Part 8: Speech Coding and Transmission,* ETS 300 175-8, Sophia Antipolis, France, September 1996.

6. ETSI, *DECT/ISDN Interworking for End System Configuration; Part 1: Interworking Specification,* ETS 300 434-1, Sophia Antipolis, France, April 1996.

7. ETSI, *Draft DECT/ISDN Interworking for Intermediate System Configuration; Interworking and Profile Specification,* ETS 300 822, Sophia Antipolis, France, January 1998.

8. ETSI, *Integrated Services Digital Network (ISDN); Digital Subscriber Signalling System No. 1 (DSS1) protocol and Signalling System No. 7 Protocol; Signalling Application for the Alpha Interface of Cordless Terminal Mobility (CTM)*, DE/SPS-05121, Sophia Antipolis, France, 1998.

9. ETSI, *Digital Enhanced Cordless Telecommunications (DECT); Common Interface (CI); Part 2: Physical Layer (PHL)*, ETS 300 175-2, Sophia Antipolis, France, September 1996.

10. ETSI, *Digital Enhanced Cordless Telecommunications (DECT); Common Interface (CI); Part 6: Identities and Addressing*, ETS 300 175-6, Sophia Antipolis, France, September 1996, with Amendment 1 of August 1997.

11. ITU-T Recommendation G.821: *Error Performance of an International Digital Connection Operating at a Bit Rate Below the Primary Rate and Forming Part of an Integrated Services Digital Network*, Geneva, 1996.

12. ETSI, *DECT Data Services Profile (DSP); Profile Overview*, ETR 185, Sophia Antipolis, France, December 1995.

13. ETSI, *A High Level Guide to the DECT Standardisation*, ETR 178, Sophia Antipolis, France, January 1997.

14. ETSI, *DECT Data Services Profile (DSP); Base Standard Including Interworking to Connectionless Networks (Service Types A and B, Class 1)*, ETS 300 435, Sophia Antipolis, France, February 1996.

15. ETSI, *DECT Data Services Profile (DSP); Generic Frame Relay Service With Mobility (Service Types A and B, Class 2)*, ETS 300 701, Sophia Antipolis, France, October 1996.

16. ETSI, *DECT Data Services Profile (DSP); Generic Data Link Service for Closed User Groups (Service Type C, Class 1)*, ETS 300 699, Sophia Antipolis, France, October 1996.

17. ETSI, *DECT Data Services Profile (DSP); Generic Data Link Service; Service Type C, Class 2*, ETS 300 651, Sophia Antipolis, France, September 1996.

18. ETSI, *Draft Digital Enhanced Cordless Telecommunications (DECT); Data Services Profile (DSP); Point-to-Point Protocol (PPP) Interworking for Internet Access and General Multi-Protocol Datagram Transport*, EN 301 240, Sophia Antipolis, France, August 1997.

19. ETSI, *DECT Data Services Profile (DSP); Multimedia Messaging Service (MMS) With Specific Provision for Facsimile Services; (Service Type F, Class 2)*, ETS 300 755, Sophia Antipolis, France, May 1997.

20. ETSI, *DECT Data Services Profile (DSP); Low Rate Messaging Service; (Service Type E, Class 2)*, ETS 300 757, Sophia Antipolis, France, April 1997.

21. ETSI, *DECT/GSM Interworking Profile (IWP); Implementation of Short Message Service, Point-to-Point and Cell Broadcast*, ETS 300 764, Sophia Antipolis, France, May 1997.

22. ETSI, *Draft Digital Enhanced Cordless Telecommunications (DECT); Data Services Profile (DSP); Isochronous Data Bearer Services With Roaming Mobility (Service Type D, Mobility Class 2)*, EN 301 238, Sophia Antipolis, France, August 1997.

23. ETSI, *Draft Digital Enhanced Cordless Telecommunications (DECT); Data Services Profile (DSP); Isochronous Data Bearer Services for Closed User Groups (Service Type D, Mobility Class 1),* EN 301 239, Sophia Antipolis, France, August 1997.

24. ETSI, *DECT Services, Facilities and Configurations for DECT in the Local Loop,* ETR 308, Sophia Antipolis, France, August 1996.

25. ETSI, *DECT Radio in the Local Loop (RLL) Access Profile (RAP); Part 1: Basic Telephony Services,* ETS 300 765-1, Sophia Antipolis, France, August 1997.

26. ETSI, *DECT Radio in the Local Loop (RLL) Access Profile (RAP); Part 2: Advanced Telephony Services,* ETS 300 765-2, Sophia Antipolis, France, January 1997.

27. ETSI, *Application of DECT Wireless Relay Stations (WRS),* ETR 246, Sophia Antipolis, France, November 1995.

28. ETSI, *DECT Wireless Relay Station (WRS),* ETS 300 700, Sophia Antipolis, France, March 1997.

29. ETSI, *DECT General Terminal Attachment Requirements: Telephony Applications,* TBR 010, Sophia Antipolis, France, January 1997.

30. ETSI, *DECT/GSM Interworking Profile (IWP); Profile Overview,* ETR 341, Sophia Antipolis, France, December 1996.

31. ETSI, *Digital Enhanced Cordless Telecommunications (DECT); Cordless Terminal Mobility (CTM); CTM Access Profile (CAP),* ETS 300 824, Sophia Antipolis, France, October 1997.

9

Regulation and Type Approval

9.1 United States Versus Europe

Regulation relevant to PWT in the United States is very different from the regulation of DECT in Europe. In general, there are more regulations in Europe than in the United States, although a lot of the background is similar, like the following:

- Both ETSI and TIA/EIA standards are optional, and compliance by a system manufacturer is dictated more by customer demand than by any other factor.
- The use of the radio spectrum is a matter that is subject to regulation on both sides of the Atlantic ocean.
- There are regulatory organizations behind both the DECT and PWT systems. In the United States, there is the FCC and its regulations, while in Europe there is the European Commission (EC) with a different set of regulations.

There are, however, some crucial differences that give rise to the different regulatory regimes:

- The need exists in the United States to ensure that systems other than PWT may safely operate in the unlicensed PCS bands, so additional regulation is needed there [1]. That is not required in Europe, because the EC has designated spectrum specifically for DECT through European Directives [2,3].

- There is separate radio spectrum for licensed systems in the United States, whereas the same band is used in Europe for all DECT applications. Also, in the United States, isochronous applications (such as speech) and asynchronous applications (such as LANs) have their own separate spectrum.
- The member states of Europe have all come from a background of having slightly different telecommunications systems, so European laws exist to ensure that certain essential aspects of telecommunication (e.g., speech services) converge over a period of many years. That convergence has long been achieved throughout the United States.

This chapter concentrates only on the regulations specific to DECT and PWT products. There are regulations in both Europe and the United States, such as safety, that must be met by DECT and PWT equipment but that are applicable to any other similar systems. These are not explicitly covered.

9.2 United States Regulation for PWT

As the DECT standards were being adopted in the United States for the PCS bands, regulations already were being drafted to permit equipment to operate in that spectrum as long as it did not interfere unduly with its peers. The detailed technical specifications for how that was to be achieved were being written by an expert body, not itself responsible for spectrum regulation. The specifications ultimately were adopted by the FCC and came into regulatory force [1] along with supporting standards [4].

9.2.1 FCC Requirements

The technical requirements of the FCC rules were summarized in Chapters 2 and 4. In addition to those technical requirements, there is a requirement that PCS devices be certified under the procedures of the FCC rules, Part 2, Subpart J [5] before being placed on the market.

Moreover, because the frequency bands now allocated to unlicensed PCS are in use by existing microwave fixed links, the FCC rules require the organization UTAM, Inc. (the unlicensed PCS ad hoc committee for 2-GHz microwave) to manage the transition of the band's use from private operational-fixed microwave service (OFS) to unlicensed PCS. The responsibility for "band clearing" includes the preparation of a funding plan for the activity, planning for the relocation of existing OFS stations, and the resolution of disputes over mutual interference. Those responsibilities will terminate when the FCC deems that the transition is complete. A requirement exists to obtain certification that a manufacturer of PCS devices is a participating member of UTAM.

9.2.2 Impact of PWT Regulations on DECT Equipment

The modifications made to the DECT standard at the lower layers, those responsible for accessing the radio spectrum, were made to fit DECT into the multisystem framework of a personal communications systems band at around 1900 MHz. Those modifications, however, go only part of the way toward allowing PWT to meet specific regulatory standards. The requirements of the FCC regulations for spectrum monitoring require apparently minor but quite significant changes to be made to the way in which the spectrum monitoring normally would work in a European DECT system. Few, if any, of the DECT-specific integrated circuits available in Europe could support the system changes needed to meet U.S. PWT regulations. Equipment designed specifically for PWT, however, should present little difficulty in being adapted for DECT.

9.2.3 Other Regulations

Apart from the radio-specific regulations for PWT, the authors know of no other regulation that is specific to PWT. For example, the special place of voice services in the European Union (EU), leading to a requirement for DECT systems providing voice service to interoperate regardless of manufacturer, does not apply in the United States. That means that compliance to the CPAP interoperability profile is optional in the United States, even though compliance to the GAP interoperability profile is mandatory in Europe. However, that does not indicate the absence of other regulation in the United States. For example, the U.S. safety standards for electrical business equipment [6], the standards for human exposure to electromagnetic radiation [7], and others still apply.

9.3 European Regulation for DECT

Unlike PWT in the United States, DECT's voice telephony service has a special place in the European regulatory regime. Traditionally, in European-type approval schemes, the verification that an equipment of a certain type meets regulatory requirements has been a matter for each national regulator to test and certify. More recently, the European scene has become less regulated and has permitted manufacturers to make their own tests (in some cases) and their own declarations of conformity. Nevertheless, the history of the separate nation states of Europe and the development of the EU as a single trading area have created a raft of special regulations that apply to DECT, to ensure its uniformity throughout the EU.

9.3.1 The Special Place of DECT Telephony

The EC and the European Council of Ministers consider voice telephony in general a particularly important service. That importance carries through into DECT. In the technical sense, by voice telephony we mean specifically the 3.1-kHz switched analog service and the ISDN 3.1-kHz telephony teleservice, both of which DECT can carry through its ADPCM-encoded 32-kbps channel.

It is considered that European telecommunications users have the right to expect special services such as voice telephony, called *justified cases*, to work properly throughout Europe. In other words, the functioning of an approved telecommunications terminal connected to the public network will always be just as users expect for at least a minimum essential set of operations.

Under the legislation that demands that special treatment for DECT (the "Telecommunications Terminal Equipment Directive" [8]), telephony via DECT (and GSM, too) is considered a separate justified case. Under the Directive, legally mandatory requirements are established to ensure the correct operation of the DECT telephony service. The resulting Common Technical Regulations, or CTRs; are based on the minimum set of technical specifications needed to meet the requirement to make, conduct, and release a call. They usually are published separately in TBR (Technical Basis for Regulation) documents and made legally binding as a CTR by a notification published in the EU's *Official Journal* (OJ). Equipment conforming to the appropriate CTRs may circulate freely throughout the EU.

Note that this Directive applies to terminal equipment, that is, terminals themselves and equipment such as PABXs that are ultimately connected to but not part of the public telephony network. It is important, therefore, to note that the CTRs do not apply to equipment within a network. A primary example is the RLL system; in most cases, all DECT RLL equipment is on the network side of the network termination point (NTP).

9.3.2 The Role of the EC, ETSI, and the National Administrations

ETSI, which publishes the DECT standards, is a statutory European standards body, but it is not responsible for either regulation or type approval. ETSI standards on their own are voluntary. Responsibility for regulation rests with the EC via its Directives, in conjunction with national authorities who implement laws to bring Directives into force, but who also retain certain national regulatory powers.

It is the EC's Approvals Committee for Terminal Equipment (ACTE) and Technical Regulations Application Committee (TRAC) that advise the EC on the implementation of CTRs and write the scope statements for the TBRs. It is ETSI, however, that writes and publishes the TBRs within its technical committees and projects.

Among the EC's committees is one that DECT manufacturers are advised to know well: the DECT Type Approval Advisory Board (DTAAB). It comes under the EC TRAC committee, and details can be found through the WWW at http://www.trac.org.uk/. The DTAAB has the following functions:

- To advise on the interpretation of the DECT standards and regulations;
- To act as a forum for independent test houses to consult each other about DECT system testing;
- To issue derogations and explanatory notes in cases where tests are discovered to be unclear or do not properly test what was intended;
- To coordinate the creation and maintenance of the complete CTR 22/GAP test suite.

The DECT CTRs are instruments that allow a manufacturer to have a DECT system tested in one place in Europe and then give it the right to sell the system anywhere. Once a product is tested to all its relevant CTRs and marked appropriately, it may circulate freely within the EU. However, being able to sell DECT equipment in any EU territory does not automatically mean it can be put into service. National regulators still retain the power to require licenses and make other national requirements under certain circumstances. In most EU states, however, those requirements take the form of class licenses for private applications, which must be met but which are not applied to individual pieces of equipment. If DECT is to be used in a public access role, then that almost certainly will be the subject of national regulation and licensing.

9.3.3 EMC

Although the CTRs are the main DECT-specific regulations that apply to DECT systems, other EU regulations that apply to DECT systems include recent EMC regulations, to meet the obligations of the EMC Directive (89/336/EEC [9], as modified by 92/31/EEC [10], 93/68/EEC [11], and 93/97/EEC [12]). The EMC Directive requires any piece of electrical or electronic equipment to comply with an appropriate EMC standard. The generic EMC standard for all radio systems is ETS 300 339 [13], which is specifically modified by ETS 300 329 [14] for DECT equipment.

Despite the existence of European harmonized regulations for DECT and other matters such as EMC, it still is necessary to gain approval for the connection of a DECT system to a network. It is not possible to cover the requirements for attachment of a DECT telephone system to a national analog PSTN or digital ISDN in this context. The attachment of ISDN terminals to a network is covered by relevant CTRs, whereas the attachment requirements for an analog line to a national PSTN are the province of local regulators.

9.3.4 Marking of DECT Products

When a DECT product is demonstrated to conform to all the relevant requirements from the EU Directives, it may be marked with a "CE mark." That mark confers on the product the "presumption of conformity" and means it may circulate freely anywhere within the EU. Specifically, if a product is demonstrated to conform with CTRs, it may be marked with a "crossed hockey stick" to demonstrate this. These marks may be affixed to the product, shown in the documentation, or printed on the product's packaging.

9.4 The European Common Technical Regulations

Four CTRs apply to basic DECT equipment: CTR 6, CTR 10, CTR 11, and CTR 22. The technical requirements of those official regulations are all contained in the related ETSI documents TBR 6 [15], TBR 10 [16], TBR 11 [17], and TBR 22 [18]. Two others, CTR 36 and CTR 40, apply to DECT/GSM systems and DECT/ISDN systems, respectively. Again, the technical contents are in TBR 36 [19] and TBR 40 [20].

9.4.1 Radio (CTR 6)

CTR 6 contains the minimum radio and PHL protocol requirements needed to ensure proper coexistence and operation of DECT radio equipment. Its intention is to impose the minimum regulation to ensure that DECT systems do not interfere with each other. As a standard, CTR 6 is technically equivalent to ETSI's type approval test standard ETS 300 176-1 [21]. The difference between the documents is purely one of legal definition.

In theory, CTRs apply only to equipment that is capable of connection to a public telecommunications network. Hence, some DECT equipment theoretically escapes the requirement to conform to CTR 6. However, the EC requires CTR 6 compliance from all DECT systems, even DECT-based wireless LANs that are not and cannot yet be connected to a public network. It is expected that this situation will be brought specifically under regulatory control by new Directives.

CTR 6 can be applied separately to handsets and bases, although testing requires a radio link to be set up. Therefore, that can be achieved only when a separate tester is available or when the applicant for CTR 6 approval supplies a corresponding part for the test.

9.4.2 Telephony (CTR 10)

CTR 10 specifies the telephony requirements for a DECT speech service. Its intention is to ensure that the speech performance you get from any combination of DECT fixed and portable parts is the same. As a standard, CTR 10 is technically equivalent to ETSI's type approval test standard ETS 300 176-2 [22]. The difference between the documents is purely one of legal definition.

Traditional telephony approval has been made on a complete telephone, which in DECT is a complete system comprising portable part and fixed part. However, the requirements in CTR 10 for the portable and fixed parts are separately specified to allow separate type approval, so meeting the objective of having the ability to mix and match DECT components in speech systems. That can be achieved only when a separate tester is available or when the applicant for CTR 10 approval supplies a corresponding part for the test.

9.4.3 Public Access Profile (CTR 11)

The DECT PAP is a set of DECT features required to allow a DECT public service to be offered by a public network operator. CTR 11 specifies the minimum standards to ensure equipment interoperation in a public access environment, that is, it ensures the interoperation of a public telephony service with a user's own DECT handset. Effectively, the PAP and CTR 11 have been replaced by the GAP and CTR 22, which cover all speech services, including public access.

9.4.4 Generic Access Profile (CTR 22)

The DECT GAP is a development of the PAP. It is intended to ensure that equipment will interoperate when any normal voice telephony service is provided, not just public telephony. In particular, it applies to residential and business applications as well as public access and is intended to bring the benefits of mix and match to all voice systems.

CTR 22 applies to all DECT systems intended for any voice application, whatever other application they carry. The standard must be applied even if a DECT system is capable of carrying 3.1-kHz voice using the built-in ADPCM coding.

It is important to note that the application of CTR 22 probably is not yet sufficient to guarantee the interoperation of DECT equipment. The DECT forum's members (see Glossary) are active in ensuring interoperability of GAP systems, and DTAAB is pressing ahead to ensure that regulatory tests are progressing toward giving a high degree of confidence.

9.4.5 DECT/GSM Interworking (CTR 36)

For equipment that claims to conform with the DECT/GSM interworking profile for connection at the GSM A-interface, conformance to CTR 36 and TBR 36 [19] is required. Strictly, the GSM A-interface is an interface within the GSM network, and so the provisions of the CTR do not apply to DECT fixed terminations. They apply only to handsets that claim such compliance.

Although not available at the time of this writing, there is expected to be a TBR 39 and a CTR 39 that will set out the essential requirements for DECT/GSM dual-mode portable parts and mobile equipment.

9.4.6 DECT/ISDN Interworking (CTR 40)

TBR 40 [20] applies to terminal equipment that is capable of connection to the ISDN and that uses DECT access. The TBR ensures air-interface interoperability between a fixed and portable part following either of the two DECT/ISDN interworking profiles (see Section 8.5), where the equipment is capable of providing 3.1-kHz voice telephony and where the fixed part is connected to the ISDN to provide ISDN services over the DECT air interface.

9.4.7 Application to the Fixed Part and Portable Part

The complex rules for the application of the CTRs are, in the end, simple to apply. The rules for fixed and portable parts not connected to GSM networks and not providing ISDN services are given here.

The basic test for the application of the DECT CTRs to a fixed part is to ask whether the fixed part is capable of connecting to the PSTN or ISDN and carrying 3.1-kHz speech. If it is capable of doing that, then all relevant CTRs apply, as illustrated in Figure 9.1(a).

If the fixed part is capable only of carrying data traffic, then only CTR 6 applies, to ensure mutual coexistence between DECT systems using the same radio spectrum. Similarly, if the fixed part is capable of carrying 3.1-kHz voice but is not capable of communication with the PSTN or ISDN, then only CTR 6 applies.

The decision on which CTRs to apply is simpler in the case of a DECT portable part. Only those DECT handsets that do not handle speech escape the application of all relevant CTRs, as illustrated in Figure 9.1(b). If the handset is capable of carrying 3.1-kHz speech, then CTRs 6, 10, and 22 apply.

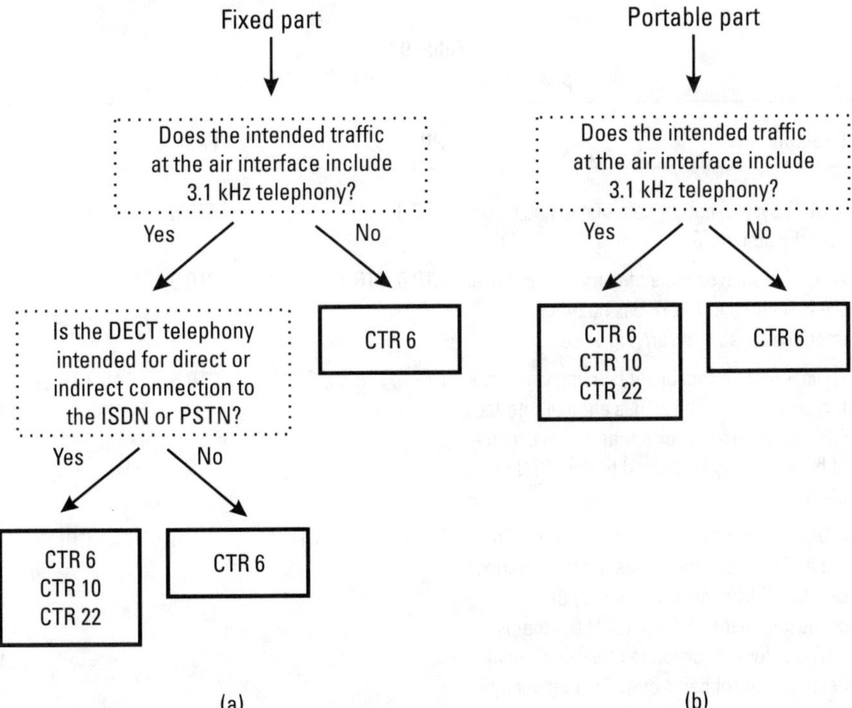

Figure 9.1 Application of the basic CTRs to (a) a fixed part and (b) a portable.

In the case of PABXs, it looks complex. You may have a system where the DECT part is integrated within a PABX, and the FP is part of the same product. Alternatively, the PABX and the DECT parts may be separate units, where the FP and the PABX are sold separately. The separate FP may be installed within the PABX, but it is not integral. In each of those cases, CTR 6, CTR 10, and CTR 22 all apply. It should be noted that connection to the European PSTN still requires equipment so connected to meet national access and voice requirements at the network interface. However, in case of clashes, CTR 10 takes preference.

9.4.8 Examples

Table 9.1 gives several specific examples of how the DECT regulations are applied to the fixed and portable parts of some complete systems.

In Table 9.1, the term "voice" is strictly a shorthand notation for "3.1-kHz ADPCM telephony."

Table 9.1
Examples of the Application of the DECT CTRs

Example	PP	FP
A DECT system not capable of carrying voice over the air interface.	CTR 6	CTR 6
A DECT system capable of carrying voice over the air interface. The FP has no other interface capable of carrying voice.	CTR 6, CTR 10, CTR 22	CTR 6
A DECT system capable of carrying voice over the air interface. The FP has another interface capable of carrying voice that is also capable of being directly connected to the PSTN or ISDN.	CTR 6, CTR 10, CTR 22	CTR 6, CTR 10, CTR 22
A DECT system capable of carrying voice over the air interface. The FP has another interface capable of carrying voice that is not directly connected to the PSTN or ISDN but that is intended for connection to other equipment itself capable of being directly or indirectly connected to the PSTN or ISDN.	CTR 6, CTR 10, CTR 22	CTR 6, CTR 10, CTR 22

References

1. Federal Communications Commission, *Unlicensed Personal Communications Service Devices, CFR 47 Part 15, Subpart D.*

2. 91/287/EEC, "Council Directive of 3 June 1991 on the Frequency Band to be Designated for the Co-Ordinated Introduction of Digital European Cordless Telecommunications (DECT) into the Community," Brussels, 1991.

3. 91/288/EEC, "Council Directive of 3 June 1991 on the Co-Ordinated Introduction of Digital European Cordless Telecommunications (DECT) into the Community," Brussels, 1991.

4. IEEE/ANSI, *American National Standard for Methods of Measurement of the Electromagnetic and Operational Compatibility of Unlicensed Personal Communications Services (U-PCS) Devices,* IEEE/ANSI C63-SC7 17-1996.

5. Federal Communications Commission, *CFR 47 Part 2, Subpart J.*

6. Underwriters Laboratories Inc., *Standard for Safety of Information Technology Equipment, Including Electrical Business Equipment,* UL 1950, February 1993.

7. IEEE, *Safety Levels With Respect to Human Exposure to Radio Frequency Electromagnetic Fields, 3 kHz to 300 GHz,* ANSI/IEEE C95.1-1992.

8. 91/263/EEC, "Council Directive of 29 April 1991 on the Approximation of the Laws of the Member States Concerning Telecommunications Terminal Equipment, Including the Mutual Recognition of Their Conformity," Brussels, April 1991.

9. 89/336/EEC, "Council Directive of 3 May 1989 on the Approximation of Laws of the Member States Relating to Electromagnetic Compatibility," *Official Journal* L139 of 23/5/89, Brussels, 1989.

10. 92/31/EEC, "Council Directive of 28 April 1992 Amending Directive 89/336/EEC on the Approximation of the Laws of the Member States Relating to Electromagnetic Compatibility," Brussels, 1992.

11. 93/68/EEC, "Council Directive of 22 July 1993 Amending Directives 87/404/EEC, 88/378/EEC, 89/106/EEC, 89/366/EEC, 89/392/EEC, 89/686/EEC, 90/384/EEC, 90/385/EEC, 90/396/EEC, 91/263/EEC, 92/42/EEC and 73/23/EEC," *Official Journal* No. L 220/1 of 30.8.93, Brussels, 1993.

12. 93/97/EEC, "Council Directive of 29 October 1993 Supplementing Directive 91/263/EEC in Respect of Satellite Earth Station Equipment," Brussels, 1993.

13. ETSI, *General Electro-Magnetic Compatibility (EMC) for Radio Equipment,* ETS 300 339, Sophia Antipolis, France, June 1997.

14. ETSI, *DECT ElectroMagnetic Compatibility (EMC) for DECT Equipment,* ETS 300 329, Sophia Antipolis, France, June 1997.

15. ETSI, *DECT General Terminal Attachment Requirements,* TBR 006, Sophia Antipolis, France, January 1997.

16. ETSI, *DECT General Terminal Attachment Requirements: Telephony Applications,* TBR 010, Sophia Antipolis, France, January 1997.

17. ETSI, *Attachment Requirements for Terminal Equipment for DECT Public Access Profile (PAP) Applications,* TBR 011, Sophia Antipolis, France, September 1994 (with amendment of March 1995).

18. ETSI, *Attachment Requirements for Terminal Equipment for DECT Generic Access Profile (GAP) Applications,* TBR 022, Sophia Antipolis, France, January 1997 (with amendment of November 1997).

19. ETSI, *Draft DECT Access to GSM Private Land Mobile Network (PLMN) for 3,1 kHz Speech Applications,* TBR 036, Sophia Antipolis, France, February 1998.

20. ETSI, *Draft Digital Enhanced Cordless Telecommunications (DECT); Integrated Services Digital Network (ISDN); Attachment Requirements for Terminal Equipment for DECT/ISDN Interworking Profile Applications,* TBR 040, Sophia Antipolis, France, September 1997.

21. ETSI, *DECT Approval Test Specification; Part 1: Radio,* ETS 300 176-1, Sophia Antipolis, France, November 1996.

22. ETSI, *DECT Approval Test Specification; Part 2: Speech,* ETS 300 176-2, Sophia Antipolis, France, November 1996.

Glossary

The final part of this book is a collected set of reference material and a means to obtain further detailed information on DECT. The Bibliography that follows adds a list of the DECT and PWT standards. This glossary comprises the following:

- A glossary of terms that are the most fundamental of the DECT-specific terms and concepts needed to understand DECT and its protocols.
- A glossary of general telecommunications terms used by DECT. These concepts are not specific to DECT but are used in the DECT system. Readers may be familiar with them from other telecommunications contexts.
- A glossary of general mobility-related terms. DECT uses a number of mobility-related terms from cellular and other telecommunications systems in specific ways.
- A list of useful organizations. Information on DECT for developers and users may be obtained from these organizations. Contact details, particularly web sites, are listed.

DECT Originally (in 1988), the Digital European Cordless Telephone. However, DECT's scope expanded soon after its conception to include non-voice communications, and it later was rechristened the Digital European Cordless Telecommunications system (thus keeping the original acronym). More recently, in response to DECT's acceptance in many non-European countries, it has become the Digital Enhanced Cordless Telecommunications system.

DECT is a digital radio transport system. It carries either a telecommunications circuit (such as a voice connection) or data packets. It is intended to connect with and extend the reach of a large number of existing communications systems, giving them cordlessness and mobility. DECT does not specify the core networks. In that way, DECT is different from digital cellular systems such as GSM, in which the network infrastructure is specified as well as the radio communications system.

PWT Personal Wireless Telecommunications (also PWT-Enhanced, or PWT-E). This is the TIA term for the U.S. version of DECT. PWT operates in unlicensed spectrum, while PWT-E (sometimes called DCT 1900) operates in licensed spectrum.

DECT-Specific Terms and Abbreviations

ARI Access rights identity. An identity that a DECT base station broadcasts on its beacon to advertise its services. In simple cases, it may derive directly from the system's hardware identity. In more interesting cases, an ARI is related to a service offered, not the equipment offering it. A portable stores the ARIs of services it can access (*see* PARK) and compares them to the ARIs on beacons it can see, to select a system to which it can attach. ARIs are classified as primary, secondary, or tertiary (PARIs, SARIs, or TARIs).

Base standards Standards containing the menu of features and procedures supporting all DECT applications, from a simple cordless telephone to a complete PABX with packet data circuits and an interface to a digital cellular network.

Bearer When the DECT protocol's MAC layer connects to another DECT MAC layer, it creates a bearer. This is a raw communication between a DECT FT and a PT. It is either a bit pipe of a given bandwidth, created by sending data packets regularly, or a packet-based bearer, in which data packets are sent on demand. A connection is built from one or more bearers.

CI Common interface. A term referring to that part of the DECT standard allowing independent DECT systems to share the same radio spectrum without mutual interference. Following the CI does not imply that the systems can interoperate. The CI is the lowest level of a DECT system.

Cluster A group of cells controlled by one MAC layer. There can be up to 255 radio cells in a cluster, but if more than that is needed in a PABX or similar multicell system, there have be two or more separate instances of the MAC layer. Often, it is convenient for an implementation to have smaller numbers of cells in a cluster anyway.

Connection What the DECT DLC layer creates out of one or more bearers. In simple cases, such as voice communication, one duplex 32-kbps bearer becomes one voice-carrying connection. In more complex cases, such as carrying ISDN for example, a connection is made up of several bearers.

CPAP Customer Premises Access Profile. The United States equivalent of the European Generic Access Profile (GAP).

DLC Data link control layer. In the DECT protocol, the DLC layer sits on top of the MAC layer and takes responsibility for ensuring error-free digital communications. This protocol layer also turns the MAC's bearers into connections suited to one of DECT's supported telecommunications services.

DLEI Data link end-point identity. Distinguishes among all the data links a DLC layer may have active simultaneously.

FP Fixed part. Comprises a fixed termination (FT) and, usually, a telecommunications network interface. In simple cases, a DECT FP is just a base station.

FT Fixed termination. The DECT-specified part of a base station. In other words, it excludes the network and the network interface that make a DECT system into a fully working communications network.

GAP Generic Access Profile. The standard that says which features and procedures from the DECT base standards are needed in a DECT system carrying voice telephony. The intention of the GAP is to ensure that voice equipment from one manufacturer operates correctly with voice equipment from another. The GAP is specific to Europe. The U.S. equivalent is the CPAP.

IE Information element. One component of a complete network layer message.

IPEI International portable equipment identity. The unique hardware identity of a piece of portable DECT equipment.

IPUI International portable user identity. The identity of the user of a DECT terminal.

IWU Interworking unit. The protocol layer that adapts the functions of a network or a user application to the interface at the top of the DECT protocol stack, that is, the top of the network (NWK) layer. For example, an IWU may take an off-hook event from a portable handset and turn it into the command that the NWK layer recognizes to set up a call. For some networks and applications (but not all), DECT specifies an IWU.

LAL Location area level. The identity of a group of cells given to a portable when it attaches to a system. By checking its LAL against the location area broadcast from the nearest DECT beacon, a portable determines if it has moved location and should inform the fixed system of its new location.

LLME Lower layer management entity. A set of procedures that sits in parallel with the PHL, MAC, DLC and NWK protocol layers. The LLME manages the operation of the protocol stack. Each layers has a communication channel to the LLME. The operation of each part of the LLME is described with each DECT protocol layer.

MAC Medium access control layer. The procedures and data flows within the DECT protocol that transmit and receive data packets using the PHL, thus creating continuous (circuit-oriented) or discontinuous (packet-oriented) bit pipes for transporting data. The bit pipes are called bearers.

MCEI MAC connection end-point identity. Distinguishes among all the connections a MAC layer may have active simultaneously.

NWK Network layer. The DECT layer that receives and processes requests from the network to which the DECT system is attached and instructs the DLC layer beneath it to create the connections needed to support the service. The network, in this case, may be either the core network connected to the base station (such as the PSTN) or the application in the portable (such as a simple telephone).

PARI Primary ARI. The identity of the primary service that a fixed system offers. A PARI is broadcast on a fixed system's beacons so that portables in the neighborhood can see if the system supports a service they recognize.

PARK Portable access rights key. The data that a portable stores when it subscribes to a particular cordless service. The PARK is actually an ARI (or a part of one) that the portable can compare with a system's broadcast ARIs to determine if the fixed system offers a service that the potable will be allowed to access (see also PLI).

PHL Physical layer. The DECT radio interface. The DECT PHL defines radio carriers, modulation, and data packet structures.

PLI PARK length indicator. The number of bits from the PARK that a portable must match with an ARI it receives to tell if the subscription which the PARK represents is for the system broadcasting the ARI.

PP Portable part. A portable termination (PT) plus a telecommunications application. One specific instance of a PP is a normal cordless telephone handset, in which the telecommunications application is a normal telephone handset interface (keyboard, display, microphone, earpiece, and ringer).

PT Portable termination. The DECT-specified component of a portable part (of which a handset is one example). It excludes the user interface (such as a keyboard, display, earpiece, microphone, and ringer), which turns the DECT PT into a fully working communications terminal.

SARI Secondary ARI. Identity of an alternative service that a fixed system provides. SARIs are broadcast on a base station's beacons to tell portables in the vicinity what secondary services are offered. They are broadcast less often than the primary ARI.

TARI Tertiary ARI. Unlike the PARI or a SARI, a TARI is not broadcast by a base station. It is available only if a portable asks the system for its list of TARI.

TPUI Temporary portable user identity. An identity usually assigned when a portable attaches to a network. A TPUI is shorter than other identities (IPUIs), so less radio bandwidth is used when paging messages are sent to portables.

General Terms and Abbreviations Used by DECT

2B+D Shorthand notation for the two 64-kbps user channels plus the 16-kbps control channel in basic rate ISDN.

ADPCM Adaptive differential pulse code modulation. A means to encode speech and in doing so reduce its bit rate to preserve radio spectrum usage. Standard digitized 64-kbps speech (PCM) is processed into 32-kbps ADPCM-encoded speech in DECT systems. The 32-kbps rate used by DECT is one of several rates defined by ITU-T Recommendation G.726 (1991), *40, 32, 24, 16 kbit/s Adaptive Differential Pulse Code Modulation (ADPCM)* (Geneva, 1990).

ARQ Automatic repeat request protocol. A system in which it is expected that data will sometimes be corrupted in transmission can use an ARQ scheme to ensure that all data transmitted are finally received without error. It does that by asking the sender to retransmit any data received with errors until the data are correctly received. A packet of data has extra bits added, which allows a data packet to be validated if it has no errors. If the data packet is received without error, the receiving side sends an acknowledgment that the packet has been correctly received. If errors are detected in a data packet, the receiving side sends a request for the transmitting side to retransmit the data. The DECT MAC layer uses ARQ to protect signaling data. The DLC layer adds a further layer of ARQ for signaling and user data.

Circuit-oriented Term that usually implies a communication with dedicated facilities where available bandwidth is guaranteed and delay is fixed. It is not the same as *connection-oriented*, since a circuit could be established over either a connection-oriented or a connectionless service.

Connectionless Connectionless communication is characterized by having no setup or teardown phase and having only a data transport phase. A successful data transfer, therefore, requires that any entity expecting to receive connectionless data must always be listening for data addressed to it and that the data must carry with it at least an explicit destination address and possibly a source address as well.

Connection-oriented Connection-oriented communication is characterized by having a connection setup phase, a data transport phase, and a connection teardown phase. In other words, an identifiable connection must exist before data may be transferred, so a receiver of data always will have been warned in advance that data transfer is about to begin. Usually, that means that the data carries an explicit or implicit connection identifier.

DCS1800 Digital cellular system at 1800 MHz. Actually, DCS1800 is functionally equivalent to GSM900, but it uses a higher radio frequency band. It is less commonly used than GSM900 and used mostly in Europe.

FDD Frequency division duplex(ing). Where portable-to-base (uplink) communications take place on a different radio frequency compared to the base-to-portable communications (the downlink).

GSM900 The global system for mobile communication at 900 MHz. Sometimes just GSM. A digital TDMA cellular system, used in Europe and several other countries, but not the United States. DECT can connect to GSM and GSM-like systems (e.g., DCS1800 and PCS1900).

ISDN Integrated services digital network. The standard digital telephony network.

Packet-oriented A communication where data are transferred in packets that are potentially not continuous. That is not the same as connectionless, since a packet-oriented service may be established over either a connection-oriented service or a connectionless service.

PABX Private automatic branch exchange. A private telephone switching system. One of DECT's main applications is as a cordless PABX.

PCS1900 The personal communication service at 1900 MHz. Actually, PCS1900 is functionally equivalent to GSM900, but it uses a higher radio frequency band. It is used in the United States.

POTS Plain ordinary/old telephone service. The service provided by the standard analog telephony network.

PSTN Public switched telephone network. Generally refers to the standard analog telephony network (but could include the ISDN).

RLL Radio (in the) local loop. The use of radio to bypass the use of copper pairs in delivering communications services to a user.

TDD Time-division duplex(ing). Also known as "ping-pong," the means by which DECT's two-way communication is effected, by repeatedly sending data first one way and then the other way, on the same radio frequency. In DECT, the cycle repeats every 10 ms.

TDMA Time-division multiple access. A system in which several digital communication channels are time compressed and sent in sequence. DECT and GSM900 are TDMA systems. DECT defines 24 slots within a 10-ms frame on each radio carrier, giving 12 two-way channels per radio carrier.

Mobility-Related Terms and Abbreviations

Cell The radio coverage area of a single DECT radio fixed part or a set of colocated radio fixed parts. Generally, a cell has a range of up to 300m, but inside a building, the coverage area is likely to be irregular and of shorter range. Outdoors, the cell may range up to 1 km or more if directional antennas are used. A single DECT cell may use any of the radio channels of the DECT system. In other words, there is no fixed radio frequency planning, and channels are used by a radio fixed part according to its own assessment of local radio usage at the time.

Handover The ability for a handset (strictly the portable part) to maintain a telecommunications session as it moves from the radio coverage area of the radio fixed part in a single cell to that of a contiguous neighboring radio cell.

HLR Home location register; also home data base (HDB). This is where the subscription and service data are held for a user of a mobile network. The data here are not usually subject to frequent change, except perhaps for the pointer to the subscriber's current VLR.

Mobility As used in this book, the means of having roaming and handover. The term is used differently elsewhere.

Roaming As used in this book, the ability to make and receive calls at any point within the radio coverage area of a wireless telecommunications system. Elsewhere it implies also the capability to receive service from more than one separate wireless system.

VLR Visitor location register; also visitor data base (VDB). The database that records the current location of a user of a mobile network. The data here are ephemeral and can change on a minute-by-minute basis.

Useful Organizations

To get documents or other information related to DECT or to attend relevant industry forum meetings, you can contact a number of organizations. Their addresses are listed here. The Internet web sites supported by the organizations here are an invaluable source of information.

ETSI European Telecommunications Standards Institute. Based in the south of France, about 20 km from Nice. ETSI documents are available by contacting ETSI directly. Some documents relating to the authentication and encryption algorithms are not generally available and are not listed in the bibliography, but they are available under nondisclosure agreement through ETSI. ETSI also administers the allocation of security and identity codes for manufacturers and operators of DECT and PWT equipment.

Postal address: ETSI, F-06921 Sophia Antipolis CEDEX, France
Office address: ETSI, 650 Route des Lucioles, Sophia Antipolis, Valbonne, France
Telephone: +33 4 92 94 42 00
Facsimile: +33 4 93 65 47 16
Web site: http://www.etsi.fr/ or http://www.etsi.org/
X.400: c=fr, a=atlas, p=etsi, s=secretariat
Internet: secretariat@etsi.fr

TIA Telecommunications Industry Association. In the United States, TIA has adopted DECT under the name of PWT, which has a different radio interface to DECT suitable for the U.S. frequency allocations. In addition, TIA publishes a Customer Premises Access Profile (CPAP) in place of the PAP and GAP. Otherwise, PWT is the same as DECT.

Web site: http://www.industry.net/tia/
Internet: tia@tia.eia.org

PWT and PWT-E documents can be obtained by contacting Global Engineering Documents in the United States.

Postal address: Global Engineering Documents, 15 Inverness Way East, Englewood, CO 80112, USA
Telephone: +1-800-854-7179 (+1-303-792-2181 internationally).
Facsimile: +1-303-397-2740
Web site: http://global.ihs.com/
Internet: global@ihs.com

In the UK, the Global Engineering Documents contact is:

Postal address: RAPIDOC®, Willoughby Road, Bracknell, Berkshire RG12 8DW, UK
Telephone: +44-1344-861666
Facsimile: +44-1344-714440
Internet: global.rapidoc@ihs.com

In France, the Global Engineering Documents contact is:

Postal address: Global Info Centre, 31-35 rue de Neuilly, 92110 Clichy, France
Telephone: +33-1-40-87-17-02
Facsimile: +33-1-40-87-07-52
Internet: global.paris@ihs.com

In Germany, the Global Engineering Documents contact is:

Postal address: Information Handling Services GmbH, Global Info Centre, Fraunhofer Str. 22, D-82152 Planegg (Martinsried), Germany
Telephone: +49-89-8952690
Facsimile: +49-89-89526999
Internet: mail@ihs.de

DECT Forum The global organization of the DECT industry, formed by leading telecommunications operators and manufacturers. One of its main objectives is to open up new markets for the technology (e.g., by addressing regulators worldwide on frequency issues). It provides a platform for the exchange of experience among operators, regulators, and standardization bodies to ensure the sustained growth and acceptance of DECT worldwide. The DECT Forum was created out of two DECT bodies, the DECT Manufacturers Forum and the DECT Operators Group (DOG), that decided to coordinate their activities and merge their organizations. As of March 1998, the DECT Forum has 42 members.

Postal address: DECT Forum Secretariat, c/o Dr. Heinz Ochsner, P. O. Box 215, CH-4503 Solothurn, Switzerland
Telephone: +41 (32) 621 7041
Facsimile: +41 (32) 621 7043
Web site: http://www.dect.ch/
Internet: heinz_ochsner@ibm.net

Bibliography

Listed here are all the standards that cover DECT and PWT, including a few important generic standards.

DECT Base Standards

The DECT base standards listed here are the lists of ingredients that go in to make up a complete DECT implementation. See Section 2.1.1 for more details of how these standards fit into the entire DECT scheme.

ETSI, *Digital Enhanced Cordless Telecommunications (DECT); Common Interface (CI); Part 1: Overview*, ETS 300 175-1, Sophia Antipolis, France, September 1996.

ETSI, *Digital Enhanced Cordless Telecommunications (DECT); Common Interface (CI); Part 2: Physical Layer (PHL)*, ETS 300 175-2, Sophia Antipolis, France, September 1996.

ETSI, *Digital Enhanced Cordless Telecommunications (DECT); Common Interface (CI); Part 3: Medium Access Control(MAC) Layer*, ETS 300 175-3, Sophia Antipolis, France, September 1996.

ETSI, *Digital Enhanced Cordless Telecommunications (DECT); Common Interface (CI); Part 4: Data Link Control (DLC) Layer*, ETS 300 175-4, Sophia Antipolis, France, September 1996.

ETSI, *Digital Enhanced Cordless Telecommunications (DECT); Common Interface (CI); Part 5: Network (NWK) Layer*, ETS 300 175-5, Sophia Antipolis, France, September 1996.

ETSI, *Digital Enhanced Cordless Telecommunications (DECT); Common Interface (CI); Part 6: Identities and Addressing*, ETS 300 175-6, Sophia Antipolis, France, September 1996 with Amendment 1 of August 1997.

ETSI, *Digital Enhanced Cordless Telecommunications (DECT); Common Interface (CI); Part 7: Security Features,* ETS 300 175-7, Sophia Antipolis, France, September 1997.

ETSI, *Digital Enhanced Cordless Telecommunications (DECT); Common Interface (CI); Part 8: Speech Coding and Transmission,* ETS 300 175-8, Sophia Antipolis, France, September 1996.

DECT Test Case Library

Listed here are the DECT test case library and the protocol implementation conformance statements (PICS) you have to fill in to record the actual features implemented and needing testing. The test case library is common whether or not you are designing a system according to an application specific profile. The PICS for specific profiles are listed with the profiles themselves.

ETSI, *DECT Common Interface (CI); Protocol Implementation Conformance Statement (PICS) proforma; Part 1: Network (NWK) Layer—Portable Radio Termination (PT),* ETS 300 476-1, Sophia Antipolis, France, September 1996.

ETSI, *DECT Common Interface (CI); Protocol Implementation Conformance Statement (PICS) proforma; Part 2: Data Link Control (DLC) Layer—Portable Radio Termination (PT),* ETS 300 476-2, Sophia Antipolis, France, September 1996.

ETSI, *DECT Common Interface (CI); Protocol Implementation Conformance Statement (PICS) proforma; Part 3: Medium Access Control (MAC) Layer—Portable Radio Termination (PT),* ETS 300 476-3, Sophia Antipolis, France, September 1996.

ETSI, *DECT Common Interface (CI); Protocol Implementation Conformance Statement (PICS) proforma; Part 4: Network (NWK) Layer—Fixed Radio Termination (FT),* ETS 300 476-4, Sophia Antipolis, France, September 1996.

ETSI, DECT *Common Interface (CI); Protocol Implementation Conformance Statement (PICS) proforma; Part 5: Data Link Control (DLC) Layer—Fixed Radio Termination (FT),* ETS 300 476-5, Sophia Antipolis, France, September 1996.

ETSI, *DECT Common Interface (CI); Protocol Implementation Conformance Statement (PICS) proforma; Part 6: Medium Access Control (MAC) Layer—Fixed Radio Termination (FT),* ETS 300 476-6, Sophia Antipolis, France, September 1996.

ETSI, *DECT Common Interface (CI); Protocol Implementation Conformance Statement (PICS) proforma; Part 7: Physical Layer,* ETS 300 476-7, Sophia Antipolis, France, September 1996.

ETSI, *DECT Common Interface (CI) Test Case Library (TCL); Part 1: Test Suite Structure (TSS) and Test Purposes (TP) for Medium Access Control (MAC) Layer,* ETS 300 497-1, Sophia Antipolis, France, February 1998.

ETSI, *DECT Common Interface (CI) Test Case Library (TCL); Part 2: Abstract Test Suite (ATS) for Medium Access Control (MAC) Layer—Portable Radio Termination (PT),* ETS 300 497-2, Sophia Antipolis, France, February 1998.

ETSI, *DECT Common Interface (CI) Test Case Library (TCL); Part 3: Abstract Test Suite (ATS) for Medium Access Control (MAC) Layer—Fixed Radio Termination (FT),* ETS 300 497-3, Sophia Antipolis, France, February 1998.

ETSI, *DECT Common Interface (CI) Test Case Library (TCL); Part 4: Test Suite Structure (TSS) and Test Purposes (TP)—Data Link Control (DLC) Layer,* ETS 300 497-4, Sophia Antipolis, France, February 1998.

ETSI, *DECT Common Interface (CI) Test Case Library (TCL); Part 5: Abstract Test Suite (ATS)—Data Link Control (DLC) Layer,* ETS 300 497-5, Sophia Antipolis, France, February 1998.

ETSI, *DECT Common Interface (CI) Test Case Library (TCL); Part 6: Test Suite Structure (TSS) and Test Purposes (TP)—Network (NWK) Layer—Portable Radio Termination (PT),* ETS 300 497-6, Sophia Antipolis, France, February 1998.

ETSI, *DECT Common Interface (CI) Test Case Library (TCL); Part 7: Abstract Test Suite (ATS) for Network (NWK) Layer—Portable Radio Termination (PT),* ETS 300 497-7, Sophia Antipolis, France, February 1998.

ETSI, *DECT Common Interface (CI) Test Case Library (TCL); Part 8: Test Suite Structure (TSS) and Test Purposes (TP)—Network (NWK) Layer—Fixed Radio Termination (FT),* ETS 300 497-8, Sophia Antipolis, France, February 1998.

ETSI, *DECT Common Interface (CI) Test Case Library (TCL); Part 9: Abstract Test Suite (ATS) for Network (NWK) Layer—Fixed Radio Termination (FT),* ETS 300 497-9, Sophia Antipolis, France, February1998.

DECT Public Access Profile

The DECT PAP was the first of the access protocols written to ensure that DECT equipment from different manufacturers intended for public access application would interwork. It is now considered obsolete, and public access applications are covered by the GAP.

ETSI, *DECT Common Interface (CI); Part 9: Public Access Profile (PAP),* ETS 300 175-9, Sophia Antipolis, France, September 1996.

ETSI, *DECT Public Access Profile (PAP) test specification; Part 1: Overview,* ETS 300 323-1, Sophia Antipolis, France, April 1994, with Amendment 1 of March 1995.

ETSI, *DECT Public Access Profile (PAP) Test Specification; Part 2: PT Abstract Test Suite (ATS),* ETS 300 323-2, Sophia Antipolis, France, June 1995.

ETSI, *DECT Public Access Profile (PAP) Test Specification; Part 3: PT PICS proforma,* ETS 300 323-3, Sophia Antipolis, France, April 1994, with Amendment 1 of March 1995.

ETSI, *DECT Public Access Profile (PAP) Test Specification; Part 4: PT PIXIT proforma,* ETS 300 323-4, Sophia Antipolis, France, April 1994.

ETSI, *DECT Public Access Profile (PAP) Test Specification; Part 5: FT Abstract Test Suite (ATS),* ETS 300 323-5, Sophia Antipolis, France, June 1995.

ETSI, *DECT Public Access Profile (PAP) Test Specification; Part 6: FT PICS proforma,* ETS 300 323-6, Sophia Antipolis, France, April 1994, with Amendment 1 of March 1995.

ETSI, *DECT Public Access Profile (PAP) Test Specification; Part 7: FT PIXIT proforma,* ETS 300 323-7, Sophia Antipolis, France, April 1994.

DECT Generic Access Profile

The DECT GAP is the current European standard for interoperation between equipment from different manufacturers. It covers residential, public, and business systems and comprises a number of standards, listed below.

ETSI, *Digital Enhanced Cordless Telecommunications (DECT); Generic Access Profile (GAP),* EN 300 444, Sophia Antipolis, France, August 1997.

ETSI, *DECT Generic Access Profile (GAP); Profile Requirement List and Profile Specific Implementation Conformance Statement (ICS) proforma; Part 1: Portable Radio Termination (PT),* ETS 300 474-1, Sophia Antipolis, France, August 1996.

ETSI, *DECT Generic Access Profile (GAP); Profile Requirement List and Profile Specific Implementation Conformance Statement (ICS) proforma; Part 2: Fixed Radio Termination (FT),* ETS 300 474-2, Sophia Antipolis, France, August 1996.

ETSI, *DECT General Access Profile (GAP); Profile Test Specification (PTS); Part 1: Summary,* ETS 300 494-1, Sophia Antipolis, France, August 1996.

ETSI, *DECT General Access Profile (GAP); Profile Test Specification (PTS); Part 2: Profile Specific Test Specification (PSTS)—Portable Radio Termination (PT),* ETS 300 494-2, Sophia Antipolis, France, August 1996 (amended December 1997).

ETSI, *DECT General Access Profile (GAP); Profile Test Specification (PTS); Part 3: Profile Specific Test Specification (PSTS)—Fixed Radio Termination (FT),* ETS 300 494-3, Sophia Antipolis, France, August 1996 (amended December 1997).

DECT Data Profiles

The DECT data profiles provide interoperation standards for a whole range of data services.

ETSI, *DECT Data Services Profile (DSP); Base Standard Including Interworking to Connectionless Networks (Service Types A and B, Class 1),* ETS 300 435, Sophia Antipolis, France, February 1996.

ETSI, *DECT Data Services Profile (DSP); Generic Frame Relay Service With Mobility (Service Types A and B, Class 2),* ETS 300 701, Sophia Antipolis, France, October 1996.

ETSI, *DECT Data Services Profile (DSP); Multimedia Messaging Service (MMS) With Specific Provision for Facsimile Services; (Service Type F, Class 2),* ETS 300 755, Sophia Antipolis, France, May 1997.

ETSI, *DECT Data Services Profile (DSP); Low Rate Messaging Service; (Service Type E, Class 2),* ETS 300 757, Sophia Antipolis, France, April 1997.

ETSI, *DECT Data Services Profile (DSP); Generic Data Link Service; Service Type C, Class 2,* ETS 300 651, Sophia Antipolis, France, September 1996.

ETSI, *DECT Data Services Profile (DSP); Generic Data Link Service for Closed User Groups (Service Type C, Class 1),* ETS 300 699, Sophia Antipolis, France, October 1996.

ETSI, *Draft Digital Enhanced Cordless Telecommunications (DECT); Data Services Profile (DSP); Isochronous Data Bearer Services With Roaming Mobility (Service Type D, Mobility Class 2),* EN 301 238, Sophia Antipolis, France, August 1997.

ETSI, *Draft Digital Enhanced Cordless Telecommunications (DECT); Data Services Profile (DSP); Isochronous Data Bearer Services for Closed User Groups (Service Type D, Mobility Class 1),* EN 301 239, Sophia Antipolis, France, August 1997.

ETSI, *Draft Digital Enhanced Cordless Telecommunications (DECT); Data Services Profile (DSP); Point-to-Point Protocol (PPP) Interworking for Internet Access and General Multi-Protocol Datagram Transport,* EN 301 240, Sophia Antipolis, France, August 1997.

DECT Approval Standards

The DECT approval standards cover three areas: EMC, radio operation, and the essential elements of interoperation with speech telephony networks.

ETSI, *DECT Approval Test Specification; Part 1: Radio,* ETS 300 176-1, Sophia Antipolis, France, November 1996.

ETSI, *DECT Approval Test Specification; Part 2: Speech,* ETS 300 176-2, Sophia Antipolis, France, November 1996.

ETSI, *DECT ElectroMagnetic Compatibility (EMC) for DECT Equipment,* ETS 300 329, Sophia Antipolis, France, June 1997.

ETSI, *General Electro-Magnetic Compatibility (EMC) for Radio Equipment,* ETS 300 339, Sophia Antipolis, France, June 1997.

DECT/GSM Interworking Standards

The DECT/GSM interworking standards provide for DECT systems to be connected to a GSM network by various means, allowing DECT handsets to use the services of the GSM network and allowing the GSM network to see a DECT handset as one of its own.

ETSI, DECT/GSM Interworking Profile (IWP); Access and Mapping (Protocol/Procedure Description for 3,1 kHz Speech Service), ETS 300 370, Sophia Antipolis, France, February 1998.

ETSI, *DECT/GSM Interworking Profile (IWP); General Description of Service Requirements; Functional Capabilities and Information Flows,* ETS 300 466, Sophia Antipolis, France, July 1996.

ETSI, *DECT Access to GSM via ISDN; General Description of Service Requirements,* ETS 300 787, Sophia Antipolis, France, July 1997.

ETSI, *DECT Access to GSM via ISDN; Functional Capabilities and Information Flows,* ETS 300 788, Sophia Antipolis, France, July 1997.

ETSI, *DECT/GSM Interworking Profile (IWP); Implementation of Facsimile Group 3,* ETS 300 792, Sophia Antipolis, France, June 1997.

ETSI, *DECT/GSM Interworking Profile (IWP); Implementation of Short Message Service, Point-to-Point and Cell Broadcast,* ETS 300 764, Sophia Antipolis, France, may 1997.

ETSI, *DECT/GSM Interworking Profile (IWP); Implementation of Bearer Services,* ETS 300 756, Sophia Antipolis, France, March 1997.

ETSI, *DECT/GSM Interworking Profile (IWP); Profile Test Specification (PTS); Profile Specific Test Specification (PSTS); Part 1: Profile Test Specification (PTS) Summary,* ETS 300 702-1, Sophia Antipolis, France, October 1996.

ETSI, *DECT/GSM Interworking Profile (IWP); Profile Test Specification (PTS); Profile Specific Test Specification (PSTS); Part 2: Portable Radio Termination (PT),* ETS 300 702-2, Sophia Antipolis, France, March 1997.

ETSI, *DECT/GSM Interworking Profile (IWP); Profile Test Specification (PTS); Profile Specific Test Specification (PSTS); Part 3: Fixed Radio Termination (FT),* ETS 300 702-3, Sophia Antipolis, France, March 1997.

ETSI, *Digital Enhanced Cordless Telecommunications/Global System for Mobile Communications (DECT/GSM) Inter-Working Profile (IWP); GSM Phase 2 Supplementary Services Implementation,* EN 300 703, Sophia Antipolis, France, March 1997.

ETSI, *DECT/GSM Interworking Profile (IWP); Profile Implementation Conformance Statement (ICS); Part 1: Portable Radio Termination (PT),* ETS 300 704-1, Sophia Antipolis, France, March 1997.

ETSI, *DECT/GSM Interworking Profile (IWP); Profile Implementation Conformance Statement (ICS); Part 2: Fixed Radio Termination (FT),* ETS 300 704-2, Sophia Antipolis, France, March 1997.

ETSI, *DECT/GSM Interworking Profile (IWP); Mobile Services Switching Centre (MSC)—Fixed Part (FP) Interconnection,* ETS 300 499, Sophia Antipolis, France, September 1996.

ETSI, *Draft Digital Enhanced Cordless Telecommunications (DECT); Global System for Mobile Communications (GSM); DECT/GSM Integration Based on Dual-Mode Terminals,* EN 301 242, Sophia Antipolis, France, August 1997.

DECT/ISDN Interworking Standards

The DECT/ISDN interworking standards either provide ISDN services directly to a DECT handset or allow a standard ISDN terminal to be connected to a socket on a DECT handset. Cordless mobility across the ISDN is also covered.

ETSI, *DECT/ISDN Interworking for End System Configuration; Part 1: Interworking Specification,* ETS 300 434-1, Sophia Antipolis, France, April 1996.

ETSI, *DECT/ISDN Interworking for End System Configuration; Part 2: Access Profile,* ETS 300 434-2, Sophia Antipolis, France, April 1996.

ETSI, *DECT/ISDN Interworking for End System Configuration; Profile Test Specification (PTS); Part 1: Summary,* ETS 300 758-1, Sophia Antipolis, France, April 1997.

ETSI, *DECT/ISDN Interworking for End System Configuration; Profile Test Specification (PTS); Part 2: Profile Specific Test Specification (PSTS) for Portable Radio Termination (PT),* ETS 300 758-2, Sophia Antipolis, France, April 1997.

ETSI, *DECT/ISDN Interworking for End System Configuration; Profile Test Specification (PTS); Part 3: Profile Specific Test Specification (PSTS) for Fixed Radio Termination (FT),* ETS 300 758-3, Sophia Antipolis, France, April 1997.

ETSI, *DECT/ISDN Interworking for End System Configuration; Profile Implementation Conformance Statement (ICS); Part 1: Portable Radio Termination (PT),* ETS 300 705-1, Sophia Antipolis, France, June 1997.

ETSI, *DECT/ISDN Interworking for End System Configuration; Profile Implementation Conformance Statement (ICS); Part 2: Fixed Radio Termination (FT),* ETS 300 705-2, Sophia Antipolis, France, June 1997.

ETSI, *Draft DECT/ISDN Interworking for Intermediate System Configuration; Interworking and Profile Specification,* ETS 300 822, Sophia Antipolis, France, January 1998.

ETSI, *Digital Enhanced Cordless Telecommunications (DECT); Integrated Services Digital Network (ISDN); DECT/ISDN Interworking for Intermediate Dystem Configuration; Profile Implementation Conformance Statement (ICS); Part 1: Portable Radio Termination (PT),* EN 301 241-1, Sophia Antipolis, France, August 1997.

ETSI, *Draft Digital Enhanced Cordless Telecommunications (DECT); Integrated Services Digital Network (ISDN); DECT/ISDN Interworking for Intermediate System Configuration; Profile Implementation Conformance Statement (ICS); Part 2: Fixed Radio Termination (FT),* EN 301 241-2, Sophia Antipolis, France, August 1997.

ETSI, *Digital Enhanced Cordless Telecommunications (DECT); Cordless Terminal Mobility (CTM); CTM Access Profile (CAP),* ETS 300 824, Sophia Antipolis, France, October 1997.

ETSI, *Integrated Services Digital Network (ISDN); Digital Subscriber Signalling System No. one (DSS1) Protocol and Signalling System No.7 Protocol; Signalling Application for the Alpha Interface of Cordless Terminal Mobility (CTM),* DE/SPS-05121, Sophia Antipolis, France, 1998.

DECT Wireless Relay Station Standard

The DECT wireless relay stations (WRS) are covered by their own standard.

ETSI, *DECT Wireless Relay Station (WRS),* ETS 300 700, Sophia Antipolis, France, March 1997.

DECT/DAM Standards

Use of a DECT-specific smart card, called a DECT Authentication Module, within a DECT portable is covered by the DECT/DAM standards.

ETSI, *DECT Authentication Module (DAM),* ETS 300 331, Sophia Antipolis, France, November 1995.

ETSI, *DECT Authentication Module (DAM); Part 1: Test Specification for DAM,* ETS 300 759-1, Sophia Antipolis, France, October 1997.

ETSI, *DECT Authentication Module (DAM); Test Specification for DAM; Part 2: Test Specification for Portable Part (PP), DAM/PP Interface,* ETS 300 759-2, Sophia Antipolis, France, 1998.

ETSI, *DECT Authentication Module (DAM); Implementation Conformance Statement (ICS) proforma Specification,* ETS 300 760, Sophia Antipolis, France, June 1997.

ETSI, *3 Volt DECT Authentication Module (DAM),* ETS 300 825, Sophia Antipolis, France, October 1997.

Radio Local Loop Standards

Systems for RLL using DECT are capable of bypassing the normal copper pair connecting a residence or other building to the local telephone exchange.

ETSI, *DECT Radio in the Local Loop (RLL) Access Profile (RAP); Part 1: Basic Telephony Services,* ETS 300 765-1, Sophia Antipolis, France, August 1997.

ETSI, *DECT Radio in the Local Loop (RLL) Access Profile (RAP); Part 2: Advanced Telephony Services,* ETS 300 765-2, Sophia Antipolis, France, January 1997.

Technical Bases for Regulation

The TBRs are the technical parts of the European regulatory CTR documents. Note that TBR 011 is now considered obsolete and has been replaced by TBR 022.

ETSI, *DECT General Terminal Attachment Requirements,* TBR 006, Sophia Antipolis, France, January 1997.

ETSI, *DECT General Terminal Attachment Requirements: Telephony Applications,* TBR 010, Sophia Antipolis, France, January 1997.

ETSI, *Attachment Requirement for Terminal Equipment for DECT Public Access Profile (PAP) Applications,* TBR 011, Sophia Antipolis, France, September 1994 (with amendment of March 1995).

ETSI, *Attachment Requirement for Terminal Equipment for DECT Generic Access Profile (GAP) Applications,* TBR 022, Sophia Antipolis, France, January 1997 (with amendment of November 1997).

ETSI, *Draft DECT access to GSM Private Land Mobile Network (PLMN) for 3,1 kHz Speech Applications,* TBR 036, Sophia Antipolis, France, February 1998.

ETSI, *Draft Digital Enhanced Cordless Telecommunications (DECT); Integrated Services Digital Network (ISDN); Attachment Requirement for Terminal Equipment for DECT/ISDN Interworking Profile Applications,* TBR 040, Sophia Antipolis, France, September 1997.

ETSI Technical Reports

There are quite a few ETSI technical reports on DECT, which are listed below. The earlier ones may be out of date, but ETR 178 is a good high-level guide to DECT standardization.

ETSI, *Digital Enhanced Cordless Telecommunications (DECT) Reference Document,* ETR 015, Sophia Antipolis, France, March 1991.

ETSI, *DECT Transmission Aspects 3,1 kHz telephony Interworking With Other Networks,* ETR 041, Sophia Antipolis, France, July 1992.

ETSI, *A Guide to DECT Features That Influence the Traffic Capacity and the Maintenance of High Radio Link Transmission Quality, Including the Results of Simulations,* ETR 042, Sophia Antipolis, France, July 1992.

ETSI, *DECT Common Interface Services And Facilities Requirement Specification,* ETR 043, Sophia Antipolis, France, July 1992.

ETSI, *DECT System Description Document,* ETR 056, Sophia Antipolis, France, July 1993.

ETSI, *DECT Wide Area Mobility Using The Global System For Mobile Communications (GSM),* ETR 159, Sophia Antipolis, France, July 1995.

ETSI, *A High Level Guide To The DECT Standardisation,* ETR 178, Sophia Antipolis, France, January 1997.

ETSI, *Conformance Testing on DECT Equipment,* ETR 183, Sophia Antipolis, France, November 1995.

ETSI, *DECT Data Services Profile (DSP); Profile Overview,* ETR 185, Sophia Antipolis, France, December 1995.

ETSI, *Application of DECT Wireless Relay Stations (WRS),* ETR 246, Sophia Antipolis, France, November 1995.

ETSI, *DECT Services, Facilities and Configurations for DECT in the Local Loop,* ETR 308, Sophia Antipolis, France, August 1996.

ETSI, *DECT Traffic Capacity And Spectrum Requirement For Multi-System And Multi-Service DECT Applications Co-Existing In A Common Frequency Band,* ETR 310, Sophia Antipolis, France, August 1996 (with corrigendum of October 1996).

ETSI, *Broadband Integrated Services Digital Network (B-ISDN); Mobile Networks Requirement on B-ISDN,* ETR 337, Sophia Antipolis, France, January 1997.

ETSI, *DECT/GSM Interworking Profile (IWP); Profile Overview,* ETR 341, Sophia Antipolis, France, December 1996.

ETSI, *Digital Enhanced Cordless Telecommunications/Global System for Mobile Communications (DECT/GSM); Integration Based on Dual-Mode Terminals,* TR 101 072, Sophia Antipolis, France, June 1997.

Other DECT Standards

Other DECT standards, not classified above, are listed here.

ETSI, *Draft Digital Video Broadcasting (DVB); Interaction channel through the Digital Enhanced Cordless Telecommunications (DECT),* EN 301 193, Sophia Antipolis, France, December 1997.

ETSI, *DECT Type Approval Advisory Board; Recommendations for the accredited conformance testing of DECT equipment,* AN DTAAB DT.04, Sophia Antipolis, France, December 1997.

EIA/TIA PWT Standards

The EIA/TIA PWT standards reproduce the ETSI DECT base standards and some application profiles, but the DECT GAP is replaced by the CPAP.

TIA/EIA, *Personal Wireless Telecommunications Interoperability Standard (PWT); Part 1: Overview*, TIA/EIA 662-1-1997.

TIA/EIA, *Personal Wireless Telecommunications Interoperability Standard (PWT); Part 2: Physical Layer (PHL)*, TIA/EIA 662-2-1997.

TIA/EIA, *Personal Wireless Telecommunications Interoperability Standard (PWT); Part 3: Medium Access Control (MAC) Layer*, TIA/EIA 662-3-1997.

TIA/EIA, *Personal Wireless Telecommunications Interoperability Standard (PWT); Part 4: Data Link Control (DLC) Layer*, TIA/EIA 662-4-1997.

TIA/EIA, *Personal Wireless Telecommunications Interoperability Standard (PWT); Network (NWK) Layer*, TIA/EIA 662-5-1997.

TIA/EIA, *Personal Wireless Telecommunications Interoperability Standard (PWT); Part 6: Identities and Addressing*, TIA/EIA 662-6-1997.

TIA/EIA, *Personal Wireless Telecommunications Interoperability Standard (PWT); Part 7: Security Features*, TIA/EIA 662-7-1997.

TIA/EIA, *Personal Wireless Telecommunications Interoperability Standard (PWT); Part 8: Speech Coding and Transmission*, TIA/EIA 662-8-1997.

TIA/EIA, *Personal Wireless Telecommunications Interoperability Standard (PWT); Part 9: Customer Premises Access Profile (CPAP)*, TIA/EIA 662-9-1997.

TIA/EIA, *Personal Wireless Telecommunications Interoperability Standard (PWT); Part 10: Approval Test Specification*, TIA/EIA 662-10-1997.

TIA/EIA, *Personal Wireless Telecommunications Interoperability Standard (PWT); CPAP Profile Test Specifications*, TIA/EIA 662-10B-1997.

TIA/EIA, *Personal Wireless Telecommunications Interoperability Standard (PWT); Part 12: Cordless Radio Fixed Part*, TIA/EIA 662-12-1997.

TIA/EIA, *Personal Wireless Telecommunications Interoperability Standard (PWT); Part 13: Data Services Access Profile*, TIA/EIA 662-13-1997.

U.S. PWT-E Standards

The U.S. PWT-E standards are based on the standards for PWT, with differences specified in the following standard:

TIA/EIA, *Personal Wireless Telecommunications—Enhanced Interoperability Standard (PWT-E)*, TIA/EIA 696-1996.

U.S. regulatory standards required by the FCC for PWT products are listed next.

Federal Communications Commission, *Unlicensed Personal Communications Service Devices, CFR 47 Part 15, Subpart D.*

IEEE/ANSI, *American National Standard for Methods of Measurement of the Electromagnetic and Operational Compatibility of Unlicensed Personal Communications Services (U-PCS) Devices*, IEEE/ANSI C63-SC7 17-1996.

About the Authors

John Phillips is a graduate of Southampton University, UK, where he read electronics and solid-state physics. After graduation in 1978, he joined STC Technology Limited to research gallium arsenide integrated circuits, their fabrication, and their applications. In 1987 in the field of radio communications, he led work to define interoperating protocols for second generation cordless telephony (CT2) and was responsible for the development of prototype systems. As secretary and later chairman of the ETSI CT2 committee, he created the specification that ETSI currently publishes. He has extensive experience in the problems of interoperation of cordless telephony products from different manufacturers and the means to cure them.

In 1991, Mr. Phillips became the manager for European wireless access standards for Bell Northern Research and later Nortel, where he is responsible for ensuring that suitable external standards and regulations exist for new personal communication and fixed access products within Europe. This role includes responsibility for the DECT standard. He and his engineers are regular attendees, contributors, and sometimes chairs at ETSI committees.

Gerard Mac Namee graduated from the University of Manchester Institute of Science and Technology, UK, in 1983 with a degree in Electronics, specializing in communications. He joined Philips Research Laboratories to work on digital cordless and cellular projects (CT2, DECT, and GSM) before becoming project leader for a collaborative European research project charged with defining the nature of third generation personal communications systems (UMTS).

Mr. Mac Namee left research and development in 1989 to join Hutchison Telecom to work on the design and management of public cordless and cellular networks. He left Hutchison in 1992 to found Personal Telecommunications Systems Ltd. He has been actively involved in European standardization programs.

Index

The Artech House Mobile Communications Series

John Walker, Series Editor

Advanced Technology for Road Transport: IVHS and ATT, Ian Catling, editor

An Introduction to GSM, Siegmund M. Redl, Matthias K. Weber, Malcolm W. Oliphant

CDMA for Wireless Personal Communications, Ramjee Prasad

CDMA RF System Engineering, Samuel C. Yang

Cellular Communications: Worldwide Market Development, Garry A. Garrard

Cellular Digital Packet Data, Muthuthamby Sreetharan, Rajiv Kumar

Cellular Mobile Systems Engineering, Saleh Faruque

Cellular Radio: Analog and Digital Systems, Asha Mehrotra

Cellular Radio: Performance Engineering, Asha Mehrotra

Cellular Radio Systems, D. M. Balston, R. C. V. Macario, editors

Digital Beamforming in Wireless Communications, John Litva, Titus Kwok-Yeung Lo

GSM System Engineering, Asha Mehrotra

Handbook of Land-Mobile Radio System Coverage, Garry C. Hess

Introduction to Wireless Local Loop, William Webb

Introduction to Radio Propagation for Fixed and Mobile Communications, John Doble

IS-136 TDMA Technology, Economics, and Services, Lawrence Harte, Adrian Smith, Charles A. Jacobs

Land-Mobile Radio System Engineering, Garry C. Hess

Low Earth Orbital Satellites for Personal Communication Networks, Abbas Jamalipour

Mobile Antenna Systems Handbook, K. Fujimoto, J. R. James

Mobile Communications in the U.S. and Europe: Regulation, Technology, and Markets, Michael Paetsch

Mobile Data Communications Systems, Peter Wong, David Britland

Mobile Information Systems, John Walker, editor

Personal Communications Networks, Alan David Hadden

Personal Wireless Communication With DECT and PWT, John Phillips, Gerard Mac Namee

RF and Microwave Circuit Design for Wireless Communications, Lawrence E. Larson, editor

Smart Highways, Smart Cars, Richard Whelan

Spread Spectrum CDMA Systems for Wireless Communications, Savo G. Glisic, Branka Vucetic

Transport in Europe, Christian Gerondeau

Understanding Cellular Radio, William Webb

Understanding GPS: Principles and Applications, Elliott D. Kaplan, editor

Vehicle Location and Navigation Systems, Yilin Zhao

Wideband CDMA for Third Generation Mobile Communications, Tero Ojanperä, Ramjee Prasad

Wireless Communications for Intelligent Transportation Systems, Scott D. Elliott, Daniel J. Dailey

Wireless Communications in Developing Countries: Cellular and Satellite Systems, Rachael E. Schwartz

Wireless Data Networking, Nathan J. Muller

Wireless: The Revolution in Personal Telecommunications, Ira Brodsky

For further information on these and other Artech House titles, including previously considered out-of-print books now available through our In-Print-Forever™ (IPF™) program, contact:

Artech House
685 Canton Street
Norwood, MA 02062
781-769-9750
Fax: 781-769-6334
Telex: 951-659
e-mail: artech@artech-house.com

Artech House
Portland House, Stag Place
London SW1E 5XA England
+44 (0) 171-973-8077
Fax: +44 (0) 171-630-0166
Telex: 951-659
e-mail: artech-uk@artech.house.com

Find us on the World Wide Web at: www.artech-house.com